COLD TOLERANCE
IN RICE CULTIVATION

COLD TOLERANCE
IN RICE CULTIVATION

P. Basuchaudhuri M.Sc. (Ag), Ph.D.
Formerly Senior Scientist
Indian Council of Agricultural Research

CRC Press
Taylor & Francis Group
Boca Raton   London   New York

CRC Press is an imprint of the
Taylor & Francis Group, an **informa** business

A SCIENCE PUBLISHERS BOOK

CRC Press
Taylor & Francis Group
6000 Broken Sound Parkway NW, Suite 300
Boca Raton, FL 33487-2742

© 2014 Copyright reserved
CRC Press is an imprint of Taylor & Francis Group, an Informa business

No claim to original U.S. Government works

Printed in the United States of America on acid-free paper

International Standard Book Number: 978-1-4822-4517-2 (Hardback)

Visit the Taylor & Francis Web site at
http://www.taylorandfrancis.com

CRC Press Web site at
http://www.crcpress.com

Science Publishers Web site at
http://www.scipub.net

Dedicated To My Mother
Late Lila Basuchaudhuri

PREFACE

Rice is the staple food for more than half of the world's population. It originated in China, and soon spread throughout the world. Today rice is cultivated predominantly in China, Japan, South-east Asia, South Asia Africa and Latin America. Whereas it is largely grown in sub-tropical and tropical hot and humid climate, it also grows in temperate climate during warm season. Rice is adapted to many environments from hill terraces and valley to coastal saline marshy lowlands. Efforts are being taken up to achieve suitable varieties for high yield in these situations. Rice can be grown round the year in many places of tropical and sub tropical humid regions. Bright sunshine, low humidity and low temperature are congenial for highest productivity.

Abiotic factors affecting rice cultivation include drought, salinity and cold. Ten percent of the total cultivated area is affected by cold weather, which adversely affects the production of rice. In 1980, the production of rice in Korea fell to half of the average due to damage by a cold spell. Rice cultivated in temperate regions is affected by prolonged cold, at various stages of growth, causing spikelet sterility and loss of production.

For many years it had been tried to cope with the problem by using varieties and management technologies but with little success.

The effect of cold on rice growth and production of rice is very complex. It adversely affects all stages of growth. It may cause chlorosis and ultimately necrosis. The maximum damage is likely to be caused at the microsporogenesis stage by the cold. It also results in incomplete and irregular panicle exertion. Lastly, the grain development is affected, causing spikelet sterility. Cold temperature during cropping period, at any stage, results in loss of yield.

To cope with the increasing population, by 2050 a sixty percent increase from the present rice yield increase is necessary, which can only be achieved by developing high yielding rice varieties that are tolerant to abiotic stress conditions. In the book, a detailed discussion on the influence of cold on rice at different stages of growth has been described with emphasis on injury, damage, recovery, physiological, biochemical, molecular and genetic mechanisms of cold tolerance. Crop improvement concepts have also been discussed. The book provides a clear understanding of cold stress in rice, to students and researchers with the hope that it will provide insight to the subject for further research so that the rice plants can be grown efficiently in cold regions with high productivity.

I am in debt to my family members for their support and cooperation. I am also thankful to the Calcutta Society for Professional Action in Development for their support.

P. Basuchaudhuri

Kolkata

CONTENTS

1

INTRODUCTION

Rice is principally a crop of tropical and semi-tropical regions, but is also grown in many cold regions of the world in late spring to summer. It is most susceptible to chilling temperatures amongst abiotic factors and experiences a greater yield loss.

Rice is a staple food for two third of the population of the world and is basic part of diet. As a cereal grain, it is the most widely consumed staple food for a large part of the world's human population, especially in Asia and the West Indies. It is the grain with 2nd highest worldwide production after maize (corn), according to the data of 2010. Since a large portion of maize crops are grown for purposes other than human consumption, rice is the most important grain with regard to human nutrition and calory intake, providing more than one fifth of the calories consumed worldwide by the human species.

Food Grain Production in the year 2011–2012

Grain	Production (million tones)
Wheat	701.5
Rice	485.6
Coarse grains	1168.7

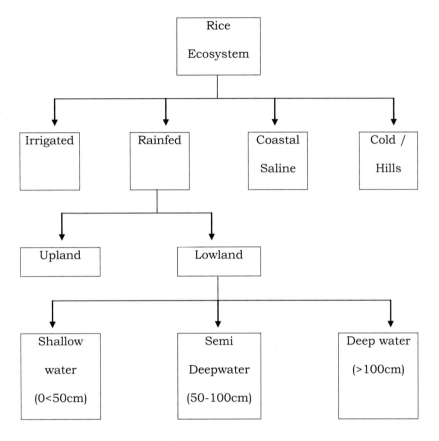

Rice can be grown in different agro-environment, in general classified as,

1. Lowland, rainfed
2. Lowland, irrigated
3. Deepwater
4. Coastal wetland
5. Upland rice

There is very little opportunity to increase the area earmarked for rice and further crop intensification is constrained by limited supply of water. Therefore, the increase in supply must mainly be met by increasing crop yields through better crop, nutrient, pest and water management and the use of germplasm and biotechnological engineering with a higher yield potential. In case of grain crops an excellent process had been introduced to have healthy plants with enriched calory and nutrient quality of grain. Also, INM, IPM and numerous improved irrigation techniques are being adopted along with application of fertilizers. Genetic enrichment and modification are being utilized and tested seriously. Among the two major types of rice viz. indica and japonica, japonica mainly grown in Japan and North China, is more tolerant to cold. Such approaches require much greater knowledge among farmers. A major challenge during the coming decade is to develop cost effective technology transfer methods to increase the ability of farmers to manage the resources at their disposal more efficiently (Fairhurst and Dobermann 2002).

Rice is a major source of food for more than 2.7 billion people and planted on nearly one-tenth of earth's available land. Of 130 million hectares of rice land, 30% is subjected to salinity problems, 20% drought and 10% low temperature at high latitude and altitude areas were well documented in Japan (Shimono et al. 2007), Korea, Northeastern and Southern China, Bangladesh, India, Nepal, and other countries (Lee 2001, Kaneda and Beachell 1974). In similar low temperature conditions severe yield losses were reported in Australia (Farrell et al. 2001), Italy and United States (Board et al. 1980).

If we look into the rice producing countries of the world we will find that about two-third of rice production is in China and India.

Top 20 rice producers by country—2010

(million metric ton)

People Republic of China	197.2
India	120.6
Indonesia	66.4
Bangladesh	49.3
Vietnam	39.3
Myanmar	33.2
Thailand	31.5
Philippines	15.7
Brazil	11.3
United States	11.0
Japan	10.6
Cambodia	8.2
Pakistan	7.2
South Korea	6.1
Madagascar	4.7
Egypt	4.3
Sri Lanka	4.3
Nepal	4.0
Nigeria	3.2
Laos	3.0

Source: FAO

The above table also indicates that Asia is the main producer of rice. However, rice is also produced in temperate regions throughout the world, though the area under rice cultivation is restricted. Other than Japan, rice is grown in United States in California and South eastern coastal belt.

Introduction

Area, yield and production of temperate rice, 1994.

Country	Yield (t/ha)	Production (X1000t)
America	5.7	16,473
USA	6.7	8,972
Chile	4.5	133
Uruguay	5.5	680
Paraguay	3.4	82
Argentina	4.3	606
S. Brazil	5.0	6000
Europe	4.2	4324
Albania	4.1	4
Bulgaria	2.9	3
France	4.5	124
Greece	8.0	120
Hungary	2.8	14
Italy	5.5	1324
Portugal	5.2	115
Romania	3.1	14
Spain	6.2	390
Turkey	5.0	200
Former USSR	3.3	2006
Near East	4.1	2920
Iran	4.3	2700
Iraq	2.4	220
North Africa	7.9	4652
Egypt	6.5	70
Morocco	6.5	70
East Asia	5.9	70536
Korea DPR	3.5	2104
Korea Rep.	6.1	7058
Japan	6.7	14976
Northern China	5.8	46400
Australia	8.3	1017

Source: FAO

It is quite evident, in temperate rice production, China and Japan are frontrunners and have evolved number of technologies to cope with the prevalent cold temperature to produce more rice. Of the 128 million ha grown in Nepal, fifty thousand ha is in Kathmandu valley and in other temperate areas are Ponlai varieties from Taiwan (Shahi et al. 1982). Kathmandu valley has about 25,000 ha of rice at an elevation of 1,365m. Of the Western Himalayan range of Jumla valley rice is grown at the highest altitude of 2,621m (Shahi and Heu 1979).

Rice production had steadily increased during the green revolution, but recently its growth has slowed down. Moreover, crop intensification during the green revolution has exerted tremendous pressures on natural resources and the environment. On the other hand, under the globalization of the world economy, rice producers are exposed to competition not only among themselves but also with the other crop producers. The future increase of rice production, therefore, requires improvement in productivity and efficiency. Innovative technologies such as hybrid rice, new plant types, and possibly transgenic rice can play an important role in raising the yield ceiling in rice production, thus increasing its productivity. Also in many countries the gaps between yield obtained and potential yield are wide. By the year 2025, the world will need about 760 million tonnes of paddy, or 35 percent more than the rice production in 1996, in order to meet the growing demand. However, available lands are mostly exploited, especially in Asia, where 90 percent of the world's rice is produced and consumed.

The productivity of cold/hill areas is very poor. The average yield is about 1.1 tonnes per ha as against the average yield of 1.9 tonnes per ha. The major problems of these areas are cold injuries, blast, drought spell, very short span of cropping seasons. Because of the rolling topography in these areas bench terracing is being followed which limits the use of fertilizers and improved agronomical practices. In these areas the crop is sometimes affected due to low temperature in early stage and sometimes at the flowering times which leads to sterility problems.

The effects of cold in the hills cause:

1. Poor germination and extended germination period.
2. Slow seedling growth, low vigor, leaf discoloration and mortality.
3. Inhibited root development, growth, and tillering and non synchronous tiller production.
4. Delayed heading and poor panicle exsertion, irregular panicle exsertion.
5. Spikelet sterility, low and irregular grain filling, and early senescence.

According to climatic conditions the cold spell affects the rice adversely in cold areas at varying growth and developmental phases even for a prolonged period in some areas. Availability of high yielding genotype is the solution no doubt, but as the life cycle of a plant is quite complex and some damages are irreversible, the solution is quite difficult. An understanding of the injury effects at different phases of the life cycle and the biological and molecular cascade of changes following the cold sensation will be elucidated vividly. As at different stages of growth and development different primordials are attained to manifest with varying functional activities, the whole system should be taken up to realize the dependent nature of functions as a whole so that a complete understanding is possible because of the fact that at a time a number of points are affected subsequently.

In this scenario, though quite a large number of research work had been carried out in the last fifty years throughout the world, many aspects await findings to come up to understand the facet totally. For example, though there are many screening techniques, because of the complex nature of the scientific problem of cold tolerance in rice, only one method is not enough to identify the tolerant variety at all stage of life cycle. It may also be mentioned that the control conditions of plant do not often match with the field condition because of diurnal variations, micro climate changes and other abiotic variations. It is expected that efforts in future especially genetic and biotechnology will pave the way for an excellent plant type to cope with the problem in general and enhance the rice production to a new high in this cold areas.

2

SEED GERMINATION

Seed germination is the initial step of plant life cycle. A small seed grows into a plant in the long run and produces many seeds. So the multiplication process goes on. In the meantime the plant gets adapted and the genomic changes in it.

Conditions essential to germination are water, air, temperature and light. For rice seed germination, quiescent or nondormant, seeds require only rehydration after release of primary dormancy. The main gas affecting seed germination is oxygen, but ethylene and to a much lesser extent, carbon dioxide can also do so. The oxygen requirement for seed germination depends greatly on other environmental factors, such as temperature, light and water potential. The higher the temperature the richer is the oxygen in the atmosphere for seed germination. However, rice seeds have been mentioned as being able to germinate in complete anoxia (Bewley et al. 2006). Except in few cases, the germination of rice seed has been known to be unaffected by light or darkness. However, the germination of rice seed is affected greatly by temperature. Cold temperature reduces seed germination starting from imbibition, activation and subsequent manifestation. Cold temperature not only reduces germination but also slows down the germination process.

Under suitable conditions, the seed absorbs water about 25% of its dry weight.

Imbibition is the rehydration of seed through the physical process of diffusion of water. According to Engels et al. (1986) the diffusion of water in white rice at 30°C is 5.2×10^{-11} m²/s. Cold temperature slows down the diffusion. The imbibition phase of germination is considered to be the most sensitive one, which leads to the escape of solutes from the seeds. In spite of the imbibition being considered the most sensitive phase, Yoshida (1981a) reported that the greatest influence of cold temperature on germination occurs in the subsequent phases of activation and growth of coleoptile and radical (Figs. 1 & 2). Cold temperature effects due to unbalanced metabolic activities retard the cell elongation and cell division (Lyons 1973).

Data indicates that the temperature range at which rice seeds can germinate is between near 0°C and 45°C. The optimal temperature however is narrower. According to Nishiyama (1976) the optimal temperature for germination of rice seeds is estimated to range from approximately 18°C to 33°C. 30°C is

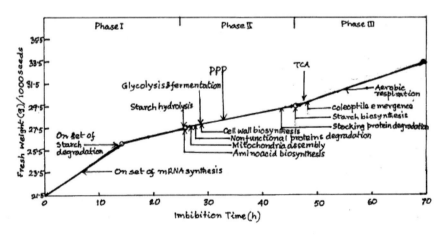

Fig. 1 Timings of metabolic activities in seed germination.

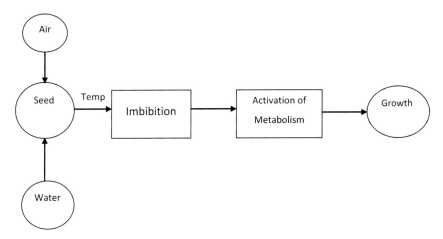

Fig. 2 Seed germination.

the standard temperature of germination of rice seeds. 10°C had been cited as the minimum critical temperature below which rice does not germinate (Yoshida 1981b). In japonica rice the germination of rice seeds is optimum at 30°C but germination was satisfactory up to 20°C. However, the germination rate decreases gradually below 20°C (Ueno and Miyoshi 2005) (Table 1).

Rice is a low temperature sensitive crop especially during seedling, tillering, panicle development and flowering stages, which can easily be injured by cold treatment. Most of the high yielding varieties cannot be used in direct sowing because

Table 1 Temperature and seed germination.

Temperature			
30°C	Optimum	Very good germination
20°C	Critical	Medium Germination
10°C	Limiting	Germination Failure

of low germination rate at low temperature. The reduction in seedling growth of rice due to low temperature is one of the major problems in tropical and sub tropical areas at high altitude as well as areas where cold mountain water is used for irrigation. In such areas, water temperature during sowing is below 15°C, whereas the optimum range for germination and early seedling growth of rice is 25–30°C. The delay in seedling emergence due to cold water, greatly increases seedling mortality and causes serious decreases in yield and increases competition with weeds. In many Asian countries, the direct seedling culture has become increasingly important in rice growing areas. Therefore, vigorous germination at low temperature is an important character for establishment of stable seedling in direct seeding culture where rice is sown directly in to flooded fields (Liang et al. 2006).

Changes in the content of starch, protein and RNA and in the activity of their hydrolases in the rice endosperm were determined during the first week of germination both in the dark and in the light, changes were generally more rapid in the dark than in the light. Oxygen uptake and RNase activity started to increase and the root protruded on the second day, followed by the coleoptile on the third day, and the primary leaf on the fourth day (Table 2).

ATP level was at a maximum on the fourth day. The activity of amylases and R enzyme increased progressively, but that of phosphorylase tended to increase during starch degradation. A new α-amylase isoenzyme band appeared during germination.

Table 2 Organ differentiation in germinating rice seed.

Organ differentiation	Days after germination
Root	Second
Coleoptile	Third
Primary leaf	Fourth

Glucose was the major product of starch degradations. Sucrose, maltose, maltotriose, raffinose and fructose were also detected. Protease activity reached a maximum on the fifth or sixth day and closely paralleled the increase in soluble aminoN and soluble protein (Palmiano and Juliano 1972).

The critical temperature for germination is 17°C. Germination and elongation decline rapidly below the critical temperature. For germination below the critical temperature some differences in metabolism is inferred.

Studies revealed that the changes in activity of enzymes in the breakdown of stored phytin, lipid, and hemicelluloses in the aleurone layer of rice seed (IR8) during the first week of the germination in the light. Enzyme activities from degermed seed such as phytase activity increased within first day of germination. The increase in activity of most other enzymes—Phosphomonoesterase, phosphodiesterase, esterase, lipase, peroxidase, catalase, beta-glucosidase and alpha and beta galactosidase closely followed the increases in protein content. Their peak activities occurred by the 5th to the 7th day. Some enzymes, such as beta-1, 3-glucanase and alpha-amylase, continued to increase in activity after the 7th day. Phytase, beta-1, 3-glucanase, and alpha amylase followed a similar sequence of production in embryo less seed halves incubated in 0.12 nM gibberellins GA3, but the production of lipase was delayed (Palmiano and Juliano 1973). In a close parallel to the developmental pattern of a-amylase activity, a rapid increase of maltase activity occured in the endosperm tissue of germinating rice seeds after about 4 days of seed imbibition. Sucrose synthetase activity show that the scutellum is the site of sucrose synthesis in germinating rice seeds, the glucose derived from starch in endosperm is transported to scutellum, where it is converted to sucrose (Fig. 3). Sucrose is

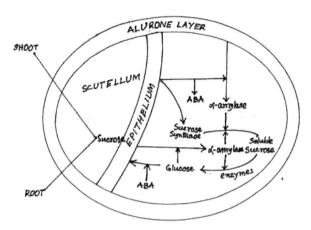

Fig. 3 Sucrose synthesis in rice scutellum.

further mobilized to the growing tissues, shoot and roots. The schematic diagram of rice sucrose formation after imbibitions can be depicted as in hereunder.

However, it is interesting to note that in the scutellum sucrose content is maximum in early period of 4–5 days after imbibitions but on the other hand glucose and fructose content increased gradually (Table 3).

The embryo plays a decisive role during seed germination. Sixty differentially, expressed proteins were identified and

Table 3 Content of sugars in scutellum of germinating rice seeds.

Time after	Sugars (mg per 100 scutella)			
Imbibition (day)	Sucrose	Maltooligo saccharide	Glucose	Fructose
4	8.3	0.01	0.31	0.50
6	7.9	0.19	0.16	0.25
8	7.8	0.22	0.34	0.55
11	7.4	0.08	0.47	0.51
14	4.1	0.08	0.69	1.17

sorted into 10 categories by their functions metabolism, oxygen-detoxifying, protein processing/degradation, stress/defense, energy and others. This suggested that there are multiple regulations in embryo during seed germination (Kim et al. 2009) (Table 4).

Table 4 Peak activities of enzymes during germination.

Enzyme	Peak activity (Approx.)
Phosphomonoesterase	5 to 7 DAG
Phosphodiesterase	"
Esterase	"
Lipase	"
Peroxidase	"
Catalase	"
Beta-glucosidase	"
Alpha-galactosidase	"
Beta-galactosidase	"
Beta-1, 3-glycanase	7 to 10 DAG
Phytase	"
ATPase	"
Pyrophosphatase	"
Protease	5 to 6 DAG
Pullulanase (R-Enzyme)	7 to 8 DAG

Aleurone is a living tissue surrounding the starchy endosperm and synthesizing key enzymes for germination. The single layer aleurone (unlike multiple layers in barley) is difficult to be separated in rice seed manually, but aleurone rich bran might be selected as substitute. 43 unique proteins with enzymatic activity (30%), signaling/regulation proteins (30%), storage protein (30%), transfer (5%) and structural (5%) were identified (Ferrari et al. 2009).

DAG = Days after germination

Dry rice seeds contain more than 17,000 stored mi RNAs. The RNA-binding proteins perform important role in keeping the stability and regulating the junctions of those long lived mi RNAs.

The presence of adequate level of the Gibberellic acid (GA3) in the seeds stimulated the synthesis, activation and secretion of hydrolytic enzymes mainly alpha-amylase, releasing reducing sugars and amino acids which are essentials for embryo growth (Mayer and Poljakoff-Mayber 1989).

During cereal seed germination, alpha-amylase in the aleurone layer plays an important role in hydrolyzing the endosperm starch into metabolisable sugars, which provide the energy for growth of roots and shoots (Akazawa and Hara-Mishimura 1985, Beck and Ziegler 1989). Previous physiological and biochemical studies have revealed that alpha-amylase expression in the aleurone layer occurs as follows. First, active gibberellins (GA) biosynthesis commences in the embryo, and the GAs transported from the embryo to the aleurone layer (Fincher 1989). Active GA s trigger the expression of alpha-amylase at the transcriptional level through the induction of a positive transactivating factor for alpha-amylase transcription (Gubler et al. 1995). The alpha-amylase is secreted from the aleurone layer into the endosperm to catalyse the hydrating reaction of stored starch (Fig. 4).

Amylase activity increased as a result of germination and on the 8th day both total activity and specific activity reached the maximum. Temperature-activity profile of *in vitro* analysis of germinating rice seed alpha amylase indicates the optimal temperature at 65°C, but at 10°C the relative activities is 20%. Thus at cold temperature the activity of alpha-amylase decreased appreciably (Shaw and Chuang 1982).

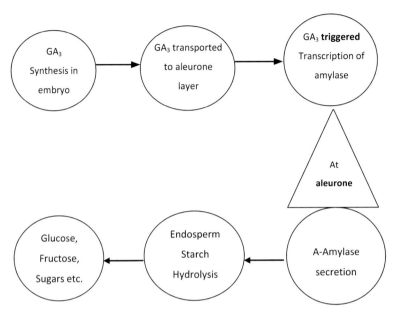

Fig. 4 GA3 regulated alpha-amylase activity.

In plants, sugars have hormone like activity and modulate nearly all fundamental processes through the entire life cycle (Smeekens 2000, Halford and Paul 2003). In cereals, the process of seedling development is divided into three stages; imbibitions, germination and seedling elongation (Thomas and Rodriguez 1994). Sugars regulate this process tightly by controlling gibberellin (GA) biosynthesis and alpha-amylase expression, which is essential for the degradation of starch to provide a nonphotosynthetic carbon source for germination and seedling development (Yu et al. 1996). After imbibing of seeds, sugars in embryo are rapidly consumed, leading to sugar depletion and subsequent activation of alpha-amylase gene expression during germination (Yu et al. 1996). Meanwhile, the embryo synthesises GA_3 that are released to aleurone cells surrounding the starchy endosperm to activate the synthesis and secretion of alpha-amylase and other hydrolases. The stored starch and other nutrients in the endosperm are digested by these hydrolases to small molecules that are taken up by the embryo to support seedling elongation (Jacobsen et al. 1995). Sugars transported to the embryo in turn repress alpha-amylase expression and GA biosynthesis in the embryo (Perata et al. 1997). Another plant hormone, abscisic acid, antagonizes GA action and represses the expression of alpha-amylase, providing a mechanism for preventing precocious germination (Jacobsen et al. 1995). Accordingly, the expression of alpha-amylase in germinating cereals is subject to multiple modes of regulation by sugars and hormones, induced by GA and sugar depletion and repressed by sugars and abscisic acid (Perata et al. 1997, Yu et al. 1996). Expression of α-amylase genes during cereal germination and seedling growth is regulated negatively by sugar in embryo and positively by GA in endosperm through the sugar response complex (SRC) and the GA response complex (GARC) respectively (Chen et al. 2006).

Cold temperature enhanced ABA synthesis which intern retards alpha-amylase activity, levels of sucrose, reducing sugars and total sugars. Suppression of RNase activity and decrease in RNA content were observed (Fig. 5).

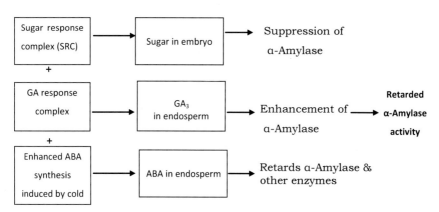

Fig. 5 Cold influence on α-amylase activity.

Unlike in non-seed tissues where ROS production is increased by ABA, ABA reduced ROS production in imbibed rice seeds, especially in the embryo region. Such reduced ROS also led to an inhibition of ASC production. GA accumulation was also suppressed by reduced ROS and ASC level, which was indicated by the inhibited repression of GA biosynthesis genes, amylase genes, and enzyme activity (Ye et al. 2011).

Pullulanase from seeds after 8 days of germination was almost equal to that from non-germinating seeds, which shows that these two enzymes are the same protein, starch debranching enzyme also known as R-enzyme. Therefore, the same pullulanase may play roles in both starch synthesis during ripening and starch degradation during germination in rice seeds.

Pullulanase in non-germinating seeds was compared with that in germinating seeds. Moreover, pullulanase from the endosperm of rice (*Oryza sativa*. Cv. Hinohikari) seeds was isolated and its properties suggest that the PI value of pullulanase from seeds after 8 days of germination was almost equal to that from non-germinating seeds. The enzyme was strongly inhibited by O-Cyclodextrin. The enzyme was not activated by thiol reagents such as dithiothreotol, 2 mercaptoethanol or glutathione. The enzyme most preferably hydrolysed pullulan and liberated only maltotriose. The pullulan hydrolysis was strongly inhibited by the substrate at a concentration higher than 0.1%. The degree of inhibition increased with an increase in the concentration of pullulan. However, the enzyme hydrolysed amylopectin, soluble starch and beta limit dextrin more rapidly as their concentrations increased. The enzyme exhibited a glucosyltransfer activity and produced an a-1, 6 linked compound of two maltotriose molecules from pullulan (Yamasaki et al. 2008).

Uptake of water by a dry seed is triphasic (Fig. 6) with a rapid initial uptake (Phase I, i.e., imbibition) followed by a plateau phase (Phase II). A further increase in water uptake occurs only after germination is completed, as the embryo axis elongates after having emerged from all seed covering layers. Phase I covered 6h from the onset of imbibition followed by a plateau (phase II) to 24h, and then the start of germination (Phase III).

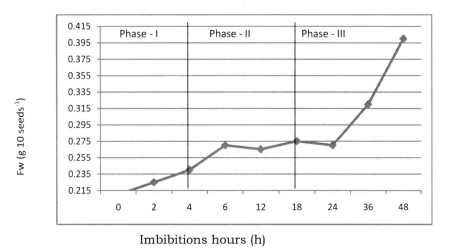

Imbibitions hours (h)

Fig. 6 Water uptake by dehusked rice seeds.

Although seed germination is a major subject in plant physiological research, there is still a long way to go to elucidate the mechanism of seed germination. Recently, functional genomic strategies have been applied to study the germination of plant seeds. A proteomic analysis of seed germination in rice (CV. 9311) by comparison of 2DE maps showed that there were 148 proteins displayed differently in the germination of rice seeds. Among the changed proteins, 63 were down regulated, 69 were upregulated (including 20 induced proteins). The down regulated proteins were mainly storage proteins, such as globulin and glutelin, and proteins associated with seed maturation such as "early embryogenesis protein "and" late embryogenesis abundant protein" and proteins related to desiccation such as "abscisic acid-induced protein" and "cold regulated protein". The degradation of storage protein happened mainly at the late stage of germination phase II (48h), while that of seed maturation and desiccation associated proteins occurred at the early stage of phase II (24h). In addition to alpha-amylase, the unregulated proteins were mainly those involved in glycolysis such as UDP glucose dehydrogenase, fructokinase, phosphoglucomutase and pyruvate decarboxylase (Yang et al. 2007).

Most of the high-yielding and high quality varieties cannot be used in direct sowing because of low germination rate at low temperature (LTG). The reduction in seedling growth of rice due to low temperature is one of the major problems in tropical and subtropical areas at high altitude as well as in areas where cold mountain water is used for irrigation. In such areas, water temperature during sowing is below 15°C whereas the optimum range for germination and early seedling growth of rice is 25 to 35°C. The delay in seeding emergence due to cold water greatly increases seedling mortality and causes serious decreases in yield and increases competition with weed. Studies with low temperature germinability with a double haploid rice (DH) population with 198 lines derived from anther culture of F1 hybrid with indica line Zenshan 97B and a perennial japonica AAV002863 was used. The germination rate in Zhenshan 97B and AAV002863 was 79.7% and 30.1%, while DH population ranged from 0 to 100% at 15°C after 6 days (Fig. 7).

Proteases hydrolyze storage proteins to provide precursors for perpetuating species. The protease activity of brown rice increased sevenfold during 7 days of germination. It is highest

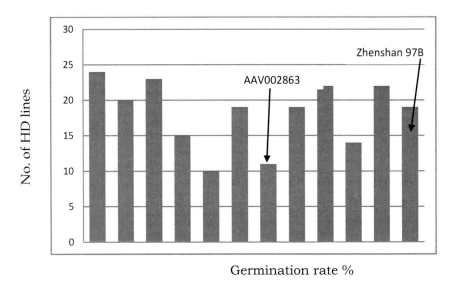

Fig. 7 Germination rate (%) in 200 rice genotypes.

on day 6. Two types of proteases known as acid and alkaline proteases are there with different functions with different properties. Ichishima (1964) had reported that most of the plant proteases are neutral or alkaline and there are few acid proteases with a pH optimum at 2–3 are widely distributed in the plant seeds and play some important physiological roles in the metabolism of seed proteins. Relative protease activity (%) is maximum around 30°C, below and above temperature decreases the activity gradually.

Activities of ATPase, acid and alkaline pyrophosphatases and phytase in the germs and endosperms of rice seed of varying germinative power were measured during their swelling and germination. Maximum ATPase activity was found in the cytoplasmic fraction. Activity of pyrophosphatase was higher in the germ and that of phytase in the endosperm. As the swelling and germination continued the enzyme activity increased (Bukhtoiarova et al. 1997). Massardo et al. (2000) demonstrated that higher metabolic rates and less oxidative damage were associated with the tolerant genotype to cold temperature during germination, as cold temperature causes a slow metabolic rate and high oxidative damages in sensitive genotypes.

On germination of rice seed, the polyamine concentration was greatest after 24hrs. and the arginine decarboxylase showed a peak after 48 hrs. (Sen et al. 1981). However, under stress condition the polyamine concentration increased over the control.

In rice seeds (cv. Tapei 309), the content of free amine and putrescine, spermidine, spermine and tyramine were higher in seed lots having a low germination frequency compared to those with high germination potential. In seeds with high germination potential, conjugates decreased drastically during germination, with an early and rapid increase in free amines. It is suggested that amines are involved in the germination process of rice seeds. It appears that amine conjugates may serve as a storage form of amines which, upon enzymic hydrolysis, could supply the cell with an additional amine reserve and influence cell division and/or cell elongation (Fig. 8).

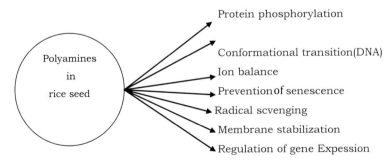

Fig. 8 Model for the regulations of polyamine metabolism in response to stress in rice.

The physiological function of polyamines under abiotic stress conditions is not clear. Polyamines are positively charged at physiological pH and are therefore able to interact with negatively charged molecules, such as nucleic acids, acidic phospholipids, proteins and cell wall components such as pectin. The multiple suggested roles of polyamines encompass involvement in protein phosphorylation, conformational transition of DNA, maintenance of ion balance, prevention of senescence, radical scavenging, membrane stabilization and regulation or gene expression (Fig. 9) by enhancing the DNA binding activity of transcription factors (Bonneau et al. 1994).

The protein of a seed, either enzymatic or storage reserves are synthesized during seed development and maturation and are deposited within membrane bound protein bodies. At the onset of germination, protein becomes degraded and the

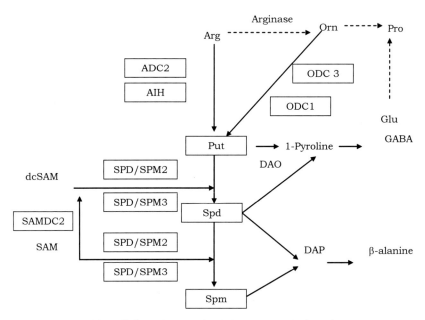

Fig. 9 Polyamine synthesis in germinating seed.

products transported to different parts of the growing plants for use in biosynthesis. The breakdown of proteins in germinating seeds and in various parts of the plant is accomplished by the activities of proteases and peptidases. Cellular protein patterns change both quantatively and qualitatively with changes in environmental conditions. Limited proteolysis and competing polypeptide degradation are a closely interacting process (Muntz 1996).

Seeds storage proteins in rice are essentially globulins which are formed in the specific storage tissue, the endosperm, during seed development. While these proteins accumulate in the respective organs, storage tissues become nitrogen, carbon sinks and acts as a nitrogen and carbon source when the protein reserves are reactivated. Within the changing sink/source relationship the formation and degradation of proteins is unique, and plays important role in manifestation of plant growth and development.

Cold stress in rice delays germination and emergence; soil temperature of below 10°C can result in complete failure of germination (Yoshida 1981b). Screening for cold tolerance based on germination and seeding growth have been attempted in rice as well (Cruz et al. 2006) and there was marked genetic variability for the traits (Satya and Saha 2010). Yoshida (1981b) studied the effect of cold stress at three phases; germination, imbibition, activation and post germination growth. The effect of cold stress was more pronounced at the phase of imbibing and this was regarded as the most sensitive phase. The exposure of seeds to cold stress during this phase has resulted in increased escape of solutes from the seeds. This has been attributed to the incomplete plasma membrane of the dry seed and the disturbance caused on its reconstruction (Cruz et al. 2004). Cold stress at this stage has been reported to target the cellular membrane and thus is the primary cause of other metabolic disorders usually observed within the cells (Lyons 1973). ABA, reduced ROS (Reactive oxygen species) production in imbibed rice seeds, specially in embryo region are known, such reduced ROS also led to an inhibition of ASC (ascorbic acid) production GA_3 (Gibberellin) accumulation was also suppressed by a reduced ROS and ASC level, which was indicated by the inhibited expression of GA_3 biosynthesis genes, amylase genes and enzyme activity (Zhang et al. 2012) (Fig. 10).

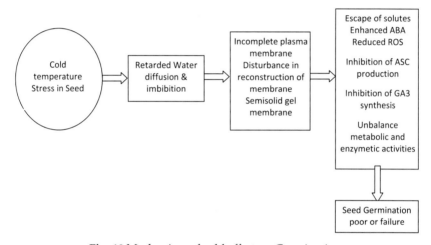

Fig. 10 Mechanism of cold effect on Germination.

3
VEGETATIVE GROWTH

Rice is a low temperature-sensitive crop especially during seedling, tillering, panicle development and flowering stages, which can be easily injured by cold treatment. Most of the high-yielding varieties do not perform well, the reduction in seedling growth of rice due to low temperature is one of the major problems in tropical and subtropical areas at high altitude as well as areas where cold mountain water is used for irrigation. In subtropical areas the summer rice also suffers from cold during early growth.

Tanaka and Yamaguchi (1969) studied the effect of temperature on the early growth of seedlings. The growth rate increased with increasing temperature over a range from 20°C to 30°C.

After seedling emergence, the root structures in young seedlings show higher weight proportions than shoot and hence, soil temperature also affects their growth and development.

Nishiyama (1977) reported that the critical minimum temperature for shoot elongation ranged from 7 to 16°C and the root elongation from 12 to 16°C. The critical minimum for elongation of both shoot and root is, hence, about 10°C. Depending on the cultivars, seed history and cultural managment practices, these critical temperatures may vary (Table 5).

Table 5 Critical temperatures in rice seedling growth.

Plant Part	Critical temperature
Root	12 to 16°C
Shoot	7 to 16°C
Shoot & Root	10°C

According to Yoshida (1973) effects of temperature on growth of IR8 in a controlled environment, in general, growth rate, relative growth rate, tillering rate increased gradually with the temperature under experiment (Figs. 11 & 12).

However, the growth efficiency (defined as the ratio of produced dry matter to the sum of produced dry matter and respiratory consumption) of rice seedlings germinated in dark was constant over the same temperature range. Sasaki and co-

Growth (relative rate %)

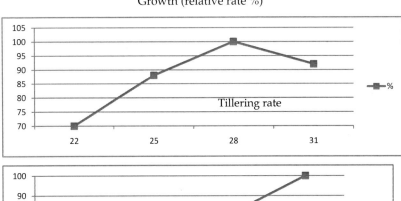

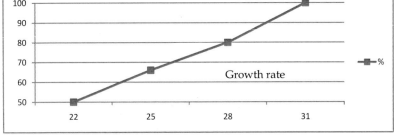

Temperature °C

Fig. 11 Tillering and seedling growth rate in rice.

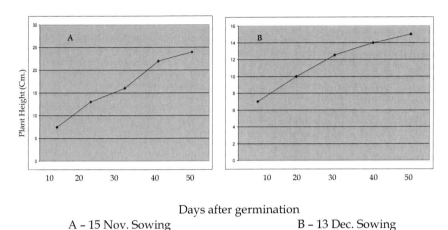

Days after germination

A – 15 Nov. Sowing   B – 13 Dec. Sowing

Fig. 12 Low temperature changes on plant height.

workers studied the relationship between germination at low temperature and subsequent early growth of rice seedlings. Correlations between germination coefficient and plant height, leaf length, dry weight and leaf number at an early stage of growth were highly significant statistically (Sasaki 1986). Positive correlation was also obtained between germination at low temperature and root development at an early stage (Sasaki and Yamazaki 1971). Development rate fluctuates as a function of air and water temperature and of photoperiod, these factors are especially important during the vegetative phase. Cold during this phase slows down rice development and lengthens growth duration. Seedling discolouration is commonly seen in indica varieties grown under low temperatures. The discolouration is usually of various degrees of yellowing of leaves and white bands on sheaths or white ring of entire leaves are observed. At tillering stage, the yellowing occurs on lower leaves. Stunting occurs at tillering stage in boro (summer) crops, also in japonica varieties grown at 9°C and 13°C (Chamura and Honma 1973).

The type of cold damage that usually occur in the hilly areas are (1) poor germination and extended germination period; (2) slow seedling growth, low vigor, leaf discolouration and mortality; (3) inhibited root-development, growth, and tillering and nonsynchronous tiller production; (4) delayed heading and poor panicle exsertion; and (5) spikelet sterility, low and irregular grain filling and early senescence (Kaw 1985). However, comparatively low temperatures were favourable for raising short but strong seedlings.

The rooting of rice seedlings occurs favourably over a range from 9°C and 33°C with an optimum at 25°C–28°C, it is severly inhibited by temperature below 16°C and above 35°C (Chamura and Honma 1973). Root number also notably decreased at 9°C–13°C. Cell division and elongation of root cells in rice seedlings, in root tip cortical cells, cell division was most active at 25°C, it decreased notably at 15°C, the elongation of the length of seminal roots at 15°C was very slow (Yamakawa and Kishikawa 1957). Shimizu (1958) showed that mitosis was normal at and above 20°C, but very slow at 15°C and ceased extirely at 10°C (Table 6).

Table 6 Influence of low temperature on number of dividing cells and cell length in rice roots at seedling stage.

Temperature (°C)	Dividing Cells (no/root tip)	Cell length (μ)
45	0	62
40	0	70
35	125	97
30	175	105
25	200	95
20	175	80
15	140	72

At later stages (3–5 weeks after sowing) temperature affects the tillering only slightly, except at the lowest temperature 12°C. Basically, higher temperatures increase the rate of emergence, and provide more tiller buds. Under low light conditions, some of the tiller buds may not develop into tillers because of a lack of carbohydrate necessary for growth. Under these conditions, low temperature may produce more tillers. When light is adequate, however, higher temperatures increase tiller number (Yoshida 1973) the effects of low air humidity and low root temperature on water uptake and growth indicated that the daily transpiration of the plants grown at low humidity was 1.5 to 2 fold higher than that at high humidity. Low root temperature (LRT) at 13°C reduced transpiration, and the extent was larger at lower humidity. LRT also reduced total dry matter production and leaf area expansion and the extent was again larger at lower humidity. These observations suggest that the suppression of plant growth by LRT is associated with water stress due to decreased water uptake ability of the root.

It is indicated, that there is a positive relationship between root oxidizing activity and dry matter production in rice seedlings. Root oxidizing activity in the cold tolerant variety was higher than in the cold-susceptible one under chilling conditions, suggesting it could be used as one of the criteria for selecting cold-tolerant rice varieties. Where root oxidizing activity fell to about 2 mg NA/2 hrs/g dryroot, dry matter production of rice seedlings stops and probably, this value was near the threshold for dry matter production (Table 7). The activity of peroxidase was also related to the root oxidizing activity under low temperature so that peroxidase can be used as another parameter for selecting cold-tolerant varieties. Content of proteins in rice roots is one of the mechanisms of rice resistance or adoption to a low temperature stress. Severe chilling stress decreased the amounts of DNA and RNA indicating a suppression of transcription process (Dai 1988).

Top growth of the rice plants after transplanting is, in general, linearly accelerated by raising average temperature from an approximately 18°C to 33°C. Above and below this range the growth decreases notably. Water temperature is

Table 7 Effect of low temperature on $\propto$–NA oxidation of rice roots (12°C).

Variety	Days after Treatment	$\propto$-NA oxidation (mg/2 hrs./g dry root)	Reduction as % of control	Control $\propto$–NA oxidation (mg/2 hrs./g dry root)
IR8	5	2.013 ± 0.111	61.4	5.210 ± 0.184
	10	1.214 ± 0.084	77.8	5.473 ± 0.171
	15	0.751 ± 0.077	86.7	5.654 ± 0.163
	20	0.414 ± 0.097	92.9	5.821 ± 0.225
Mean		1.098		5.539

much more significant than air temperature for early growth and development of low land rice (Hoshino et al. 1969). This is because the growing points of the plants are under water.

Shoot elongation at early seedling stage ranges from 7°C to 16°C and that for root elongation from 12°C to 16°C (Nishiyama 1977). Hence about 10°C may be considered as the critical minimum for elongation of both shoot and root (Table 8). The optimum temperature for cell division of the radicle tip is 25°C and that for cell enlargement 30°C. Elongation of the radicle stops below 15°C.

Favourable temperature range for physiological processes of rice plants is from 15°C–33°C. However, the growth of roots has an optimal temperature around 25°C and lower temperatures

Table 8 Growth of rice seedling.

Variety	Seedling Length (cm.)		No. of Leaves		Seedling dry weight (g)	
	Cont.	LT	Cont.	LT	Cont.	LT
IR58025A	27.7	13.8	3.56	3.00	0.027	0.017
Giza181A	18.7	13.6	3.36	3.00	0.024	0.022
IR58025A X Giza181A	24.5	11.2	3.50	3.00	0.023	0.014
Mean	23.6	16.2	3.47	3.00	0.025	0.018

(after Abdelkhalik et al. 2010)

are favourable to raising strong seedlings. There is a controversy about the effect on tillering in relation to temperature within the range. A number of researchers reported that the number of tillers increased with decreasing temperature. Among them Matsushima et al. (1966) noted that low temperature is not favourable to the elongation of tillers. Yoshida (1973) suggested that tillering of rice plant should be considered in terms of interaction between light intensity, temperature, and carbohydrate metabolism. Relatively low temperature (19°C) in the early stage of development had an important effect on the sprouting of tillers, and this effect remained as an after effect of temperature during the subsequent stage of development determining tillering pattern.

The above findings highlight the importance of atmospheric temperature as an environmental factor predictor variable in determining tillering and yield.

Atmospheric temperature gave the highest direct contribution among the environmental factors to tiller production and yield (Garba et al. 2007).

In a recent study it has been noted that low temperature leads to a low tiller occurring. The effect of low temperature was different among the different varieties of rice. The number of tillers of the rice treated in low temperature for 3 to 6 days was significantly lower (Li-Zhil et al. 2009).

According to Pathak (1991) rice plants are most sensitive at vegetative stage to low water temperature during active tillering stage which is common at higher altitudes where the water for irrigation comes from the melting snow and glaciers. Under low temperature the leaf area is reduced due to production of fewer and smaller leaves with impaired photosynthetic activity (Figs. 13 & 14).

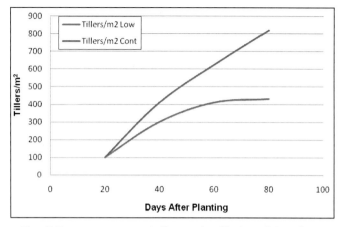

Fig. 13 Low temperature influence in tillering of rice plant.

Color image of this figure appears in the color plate section at the end of the book.

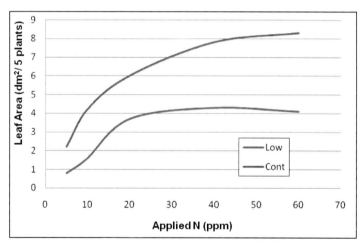

Fig. 14 Leaf area changes in rice seedlings under varying levels of nitrogen.

Color image of this figure appears in the color plate section at the end of the book.

On the other hand, the net assimilation rate was not affected by low humidity and LRT, and the water use efficiency was larger for LRT (Kuwagata et al. 2012) (Table 9 & Fig. 15).

Table 9 Effect of low root temperature on the net assimilation rate (gm⁻²d⁻¹) of rice plant under different humidity and root temperature.

Before LRT 25°C		During LRT Period	
		25°C	13°C
Humid	21.7 ± 1.5	15.9 ± 0.6	16.3 ± 0.8
Dry	21.9 ± 1.5	15.3 ± 1.5	14.8 ± 0.3

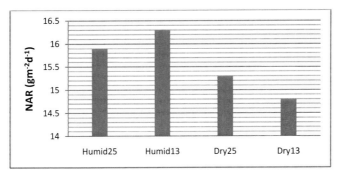

Fig. 15 Changes of net assimilation rate with humidity and low temperature.

Thus, from the above findings it is clear that dry matter production, net assimilation rate as well as stomatal conductance were reduced severely under dry cold climate. Humid cold climate had some sparing effects on the vegetative growth of the rice plants (Kuwagata et al. 2012) (Figs. 16, 17 & 18).

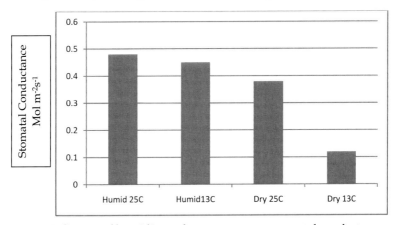

Fig. 16 Influence of humidity and temperature on stomatal conductance (after Kuwagata et al. 2012).

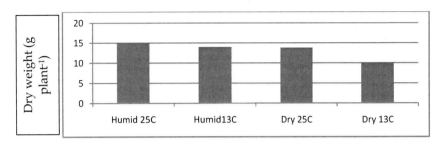

Fig. 17 Changes of dry weight under varying humidity and root temperature.

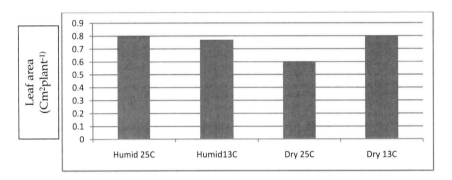

Fig. 18 Changes of leaf area under varying humidity and root temperature
(at 55 days after planting).

Table 10 Nitrate and ammonium uptake (μ moles gfw^{-1}) as influenced by
temperature in growing rice seedlings.

Hours	NO$_3$-uptake		NH$_4$-uptake	
	15°C	25°C	15°C	25°C
1	2.0	2.5	5.0	7.0
2	3.0	5.0	11.0	11.0
4	3.5	15.0	15.0	20.0
5	5.0	22.0	19.0	28.0
7	5.0	33.0	20.0	35.0

In a water culture experiment (Table 10) the rice seedlings showed no appreciable nitrate uptake at temperature below 15°C, whereas appreciable ammonium uptake occurred at 15°C, although lowered passive absorption and complete inhibition of the rapid metabolism dependent absorption phase were observed at 5°C (Sasakawa and Yamamoto 1978).

Although attention has focused largely on the consequences of moderate and prolonged cold on plant performance, brief unseasonal cold can also have a damaging effect on growth and survival of the plants, may be at risk. In some regions sudden, short cold spells may still occur during generally warmer weather that can result in checks to growth. Rice is a plant of tropical origin that is susceptible to damage by low temperature below 15°C. Cultivars of rice vary in their cold susceptibility with japonica more cold tolerant than indica cultivars. In seedling stage, the synthesis of several new proteins, alternations to lipid saturation and chlorosis, where the newly emerging leaves lack chlorophyll (Kazemitabar et al. 2003). In many plants, chlorophyll biosynthesis in chill (7°C) is affected by a sharp reduction. Inhibition of chlorophyll biosynthesis was partly due to impairment of 5-aminolevulinic acid biosynthesis. Protochlorophyllide synthesis in chill was inhibited because of defects in protoporphyrin IX synthesis, Mg-chelatase and Mg-protoporphyrin IX, monoesterase cyclase (Tewari and Tripathy 1998).

Plant mitchondria have two oxidizing pathways, the cytochrome pathway (CP) and the alternative pathway. The latter consists of one enzyme, the alternative oxidase (AOX). AOX is not coupled to H^+ translocation, and therefore ATP production, and has the potential to catalyze wasteful respiration in higher plant mitochondria. AOX is thought to prevent production of reactive oxygen species (ROS) in the respiratory chain, especially under stress conditions, by helping to prevent over-reduction of ubiquinone pool. Many abiotic stresses, including sudden exposure to low temperature induce synthesis of AOX. The increase in AOX capacity at low temperature is often associated with *de novo* synthesis of AOX protein.

Some plants have the ability to maintain similar respiratory rates when grown at different temperatures. This phenomenon is referred to as respiratory homeostasis. Using rice cultivars with different degrees of respiratory homeostasis, it was demonstrated that high homeostasis cultivars maintain shoot and root growth at low temperatures (Kurimoto et al. 2004).

Four indica and five japonica varieties of rice (*Oryza sativa* L.) had been found with their differences in photosynthetic activity and dark respiration rate influenced by leaf nitrogen levels and temperatures. The photosynthetic rate of single leaf showed correlation with total nitrogen and soluble protein content When compared at the same level of leaf nitrogen or soluble protein content, the four indica varieties and one japonica variety, Tainung 67, which have some indica genes derived from one of its parents, showed higher photosynthetic rates than the remaining japonica varieties. When the leaf temperature rose from 20 to 30°C, the photosnthetic rates increased by 18 to 41% whereas respiratory rates increased by 100 to 150%. These increasing rates in response to temperature were higher in the japonica than in the indica varieties. In this respect Tainung 67 showed the same behavior as of other four japonica varieties (Weng and Chen 1987).

Tsunoda and Khan (1968) found differences in the photosynthetic tissue of indica and japonica types. The indica types have arranged chlorenchyma cells sparsely in the mesophyll, the adaxial surface of the mesophyll is flatter. The japonica types have chlorenchyma cells more compactly arranged and the adaxial surface of the mesophyll is very wavy. However, these differences were found only at the seedling stage and not at the later stage of growth. It can be interpreted that since light is not limiting at the seedling stage, these differences may be an adaptation directly concerned with photosynthesis, rather than adaptation developed as a result of temperature. The wavy, compact mesophyll absorb and consume heat better during seedling stage. Structurally, the japonica plants are more thermodynamically adapted to low temperature at seedling stage.

The four leaf stage seedlings of two rice varieties were grown in combinations of 5 levels of nitrogen (N) (from 5 to 80 ppm) and three temperature regimes (day/night temperature of 22/17, 27/22 and 32/27°C). It was found that the photosynthetic rate was best expressed as leaf area basis. With this unit the photosynthetic rate of single leaf was closely related with chlorophyll content. There was linear relation between total N content (and protein) and chlorophyll content, indicating that photosynthesis per unit leaf area was closely correlated with nitrogen assimilation in leaves of the rice plants. Temperature and N level both affected photosynthesis of the rice leaves. In a tiller, the photosynthetic rate of leaves at different node positions was influenced by temperature and N level. At low temperature (22/17°C) there was no significant difference of photosynthesis among leaves at different node positions under different N-treatments. However, when plants grew at higher temperatures, lower photosynthetic rate was found at lower positioned leaves under low level of N, but not under high level of N. The results indicate that photosynthesis is under the control of N uptake and metabolism of rice plants.

Total soluble protein and amino-N increased with the increase in N concentration. There was higher N-content (both soluble protein and amino nitrogen) in the plants grown at lower temperature as compared with plants grown at higher temperature, particularly with plants grown at higher level of N, indicating the dilution effect of N by plant growth. Ribulose bisphosphate carboxylase and cytochrome C-oxidase activities increased with the increase in N level. However, the enzymic activities were low in all treatments of plants grown at low temperature. Nitrate and nitrite reductase activities were enhanced by N concentration. No detectable activity of nitrate reductase was found in rice roots grown at N levels below 20 ppm N (Shieh and Liao 1987).

Many factors, including exogenous carbon metabolites influence the expression of enzymes involved in nitrogen metabolism, but little is known about the effect of low temperature on the expression of those enzymes. It appears that rice seedlings grown at optimal temperature (30°C) were subjected to low temperature (20°C) stress and several key enzymes involved in ammonium assimilation and carbon metabolism, including glutamine synthetase, NADP-dependent isocitrate dehydrogenase and NADH-dependent glutamate dehydrogenase for rice roots.

The nutrient uptake is reduced due to restricted root growth, poor tillering and leaf discolouration. The early vegetative growth stage of the plant life cycle is vulnerable to cold stress marked injuries that have been observed on rice seedlings planted in early spring in temperate and subtropical environment (Andaya et al. 2003). However, the degree of injury due to cold stress varies with duration of exposure, variety and stage of development. Both root and shoot development have been shown to be very sensitive to cold stress at seedling stage. In poorly developed root system, absorption and translocation of nutrients and water is hindered affecting shoot development. Moreover, cold stress has been shown to arrest leaf growth by extending the duration of meristematic cycles (Rymen et al. 2007). It reduces root hydraulic conductance resulting in low leaf water and larger potential and ultimately reduces growth at once. This becomes irreversible, it ends up with cell death.

Nevertheless, brief exposure of seedlings to short duration of chilling temperature may reduce leaf number and plant height and once the stress is over, the plants quickly recover and resume normal growth (Majora et al. 1982).

In normal environmental condition, it may be revealed that rice plants grown under proper fertilization, gradually increase the uptake of nutrients and are correlated positively with the dry matter production. Nutrients thus observed are translocated to the developing plant parts to maximize the growth during vegetative stage. Thus, nutrients taken up actively by root system are distributed to the parts of the plant, i.e., partitioning differentiated amongst the different plant organs. As the vegetative growth is almost linear, the uptake of nutrients are usually gradually increasing, however, the concentration of different nutrients may vary during the growth depending on variety, stage of growth, abiotic conditions as well as availability of nutrients.

Cold temperature of irrigation water reduced rice shoot and root dry weight and plant height significantly. Under low temperature stress nitrogen was a major rice growth determinant. Increased shoot concentrations of both phosphorus and zinc alleviated the low temperature stress. The uptake of nitrogen, phosphorus, potassium and zinc reduced significantly at low temperature (16.5°C) with the strongest effect being noticed for nitrogen, followed by phosphorus, potassium and zinc. Application of nitrogen, phosphorus, potassium and zinc increased their uptake in rice shoots. Nitrogen and potassium had synergistic effect on their uptake. Responses to phosphorus and zinc application were well marked at low temperature than nitrogen and potassium application (Zia et al. 1994).

Soil temperature below 68°F affect nutrient uptake adversely and therefore, retard growth, yield and maturity. Soil temperature is also a major factor in affecting the availability of zinc. Larger corrective applications of zinc are required in cool seasons. Phosphorus-induced zinc deficiencies do not appear to be a significant problem in rice. Cold soil reduced phosphorus uptake, as well as, most other chemical and biological activities in the soil. Optimum soil temperature for potassium uptake is 60–80°F. Potassium uptake is reduced at low soil temperature. Low temperature retards the respiratory rate and reduces nutrient uptake.

Takahashi et al. (1954) showed that the depression of mineral absorption came in the order magnesium > manganese > silicon > calcium > nitrogen > phosphorus. The uptake of potassium seemed to be enhanced by low temperature. Fujiwara and Ishida (1963) found that the absorption of manganese was greatly enhanced by low temperature but drop below normal when low temperature treatment was extended. Fujiwara and Ishida (1963) reported that the inhibition of absorption are equally severe at 17°C for phosphorus, potassium, nitrogen, iron, silicon, calcium and magnesium for the first 2 weeks after transplantation. At tillering stage, severe inhibition occurred for phosphorus, potassium and calcium at 17°C, the absorption of manganese was accelerated. Low temperature during early growth stages causes yellowing of leaves and stunting. The limited supply of photosynthates and nutrient elements coupled with the direct effect of low temperature on the respiration rate in growing parts and the reduced efficiency, all combine to reduce the growth of the growing organs (Murty 1980). The translocation of the carbohydrates to the growing seedling is also impaired (Fig. 19).

CO_2 assimilation underpins plant productivity and is therefore central to any analysis of the response of plants to a change in temperature. Photosynthetic characteristics in rice leaves after treatment with low temperature (15°C) and high

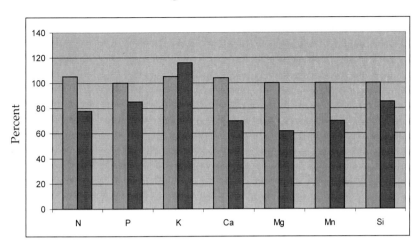

Fig. 19 Effect of low temperature on nutrient uptake in rice (Takahashi et al. 1954).

Color image of this figure appears in the color plate section at the end of the book.

irradiance (1,500 μ molquanta m^{-2} s^{-1}) decreases in quantum efficiencies in PS II (ϕ PS II) and PSI (ϕ PS I) and in the rate of CO_2 assimilation were observed with a decrease in the maximal quantum efficiency of PS II (F$_v$/Fm) by simultaneous measurements of chlorophyll fluorescence, P700$^+$ absorbance and gas exchange. The decreases in ϕ PS II were most highly correlated with those in CO_2 assimilation. Although the initial and total activities of ribulose-1, 5-biphosphate carbocylase/oxygenase decreased slightly, the maximal activity of Rubisco remained almost constant (Hirotsu et al. 2005). Herath and Ormrod (1965) reported that water temperature affected the number and size of stomata of rice seedlings, at a lower temperature (16°C), the number of stomata was smaller and their size was larger than those at higher temperature (24°C and 32°C). The temperature responses of photosynthetic assimilation rate growth were examine in rice grown hydroponically, under varying day/night temperature regimes. Irrespective of growth temperature the maximal rates of photosynthetic assimilation rate was found to be at 30–35°C in rice. However, photosynthetic assimilation rate measured at the growth temperature remained almost constant irrespective of temperature. Biomass production and relative growth rate was greatest in rice grown at 30°C/24°C. The net assimilation rate in rice decreased at low temperature (19°C/16°C) though there is no much differences in leaf area ratio. The nitrogen use efficiency (NUE) for growth rate (GR) estimated by dividing net assimilation rate (NAR) by nitrogen content correlated with GR and with biomass production (Nagai and Makino 2009).

In a tiller, the photosynthetic rate of leaves at different node positions was influenced by temperature and nitrogen level. At low temperature (22°C/17°C). There was no significant difference of photosynthesis among leaves at different node positions under different nitrogen treatments. However, when plants grew at higher temperatures lower photosynthetic rate was found at lower positioned leaves under low level of nitrogen, but not under higher level of nitrogen. There was higher nitrogen content (both soluble protein and total amino nitrogen) in plants grown at lower temperature as compared with those of higher temperature. Leaf nitrogen content is usually positively correlated with leaf photosynthetic rate. The enzyme activities (RuBPCase) and cytochrome c oxidase were low under all nitrogen treatments of plants grown at low temperature (Shieh and Liao 1987) (Fig. 20).

When a cold-tolerant cultivar Xiangnuo 1 and a cold sensitive cultivar IR50 was exposed to chilling for 2 days, the photosynthetic rates declined dramatically by 48.7% and 67.5% respectively in seedlings. Chlorophyll fluorescence measurement indicated that the reaction centres and antenna

Vegetative Growth

RuBPCase

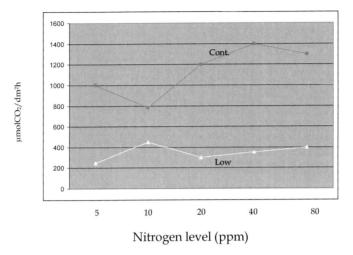

Nitrogen level (ppm)

Cytochrome C oxidase

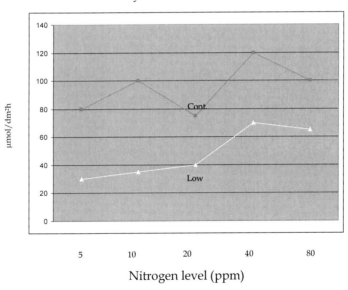

Nitrogen level (ppm)

Fig. 20 RuBPCase and Cytochrome C oxidase changes with temperature and nitrogen.

Color image of this figure appears in the color plate section at the end of the book.

systems in IR 50 were damaged severely by chilling, which led to lower photosynthetic rate (Guo-Li and Zhen-fei 2005). Leaf photosynthesis was reduced from 8.1 to 5.2 percent along with each 1°C reduction of water temperature. Stomatal conductance showed a similar response to photosynthesis and demonstrated that the differences of reduction in stomatal conductance for each degree centigrade or more was 2.6 and 3.7 percent.

However, the daily transpiration rate per plant was greater in cold tolerant cultivars compared with sensitive ones (Dai et al. 1993) due to wide stomatal opening.

Seminal rice (*Oryza sativa* L. CV. Taichung native 1) roots were grown *in vitro* to investigate the relationships among polyamine biosynthesis, root growth and chilling tolerance. At 25°C, the level of free putrescine, the activities of arginine decarboxylase (ADC) and ornithine decarboxylase (OX) increased as growth progressed. While the levels of free spermidine/spermine and the activity of S-adensosylmethionine decarboxylase (SAMDC) decreased. Exogenously applied putrescine, ranging from 0.01 to 1 mM, enhanced the elongation of roots grown at 25°C whereas application of spermidine or spermine inhibited root elongation, O-Difluoromethylarginine (DFMA) at 5 µM or α-Difluoromethyl ornithine (DFMO) at 10 µM inhibited the increase in root length and the levels of free putrescine at 25°C, these effects were reversed by the addition of 1 mM putrescine. Roots exposed to 5°C ceased growth and lost their re growth ability after 9 days of chilling. The level of free putrescine and the activity of ADC in chilled roots increased before returning to normal at day 3 and then decreased to a plateau after 9 days. The levels of free spermidine and spermine increased after 9 days. When putrescine was applied at concentrations greater than 0.1mM, the chilled roots partially recovered their re-growth ability. Contrary to DFMO (10 µM) DFMA (5 µM) inhibited both the chilling induced free putrescine. These suggest that polyanines are related to growth of rice roots cultured *in vitro* (Lee 1997).

The nature and mechanism of resistance of rice plants to chilling stress the effects of low temperature treatment (8°C) on the photosynthetic rate and some important compounds forming redox cycles were related. The varieties of rice used are two japonica rice varieties, i.e., Taipei309 and Wuyujing, three indica rice varieties, i.e., IR64, Pusa and CA 212 and one intermediate type, i.e., Shanyou 63. Three types of varieties were compared. The light intensity -photosynthesis curves, CO_2-photosynthesis curves, primary photochemical efficiency (Fv/Fm), active oxygen species (AOS) (O_2–H_2O_2), glutathione (both oxidized and reduced forms) and ascorbate contents in their six-week old seedlings were measured before and after chilling treatment. The results showed that relative to the rice varieties chilling tolerance such as Taipei 309 and Wuyujing, the sensitive ones indica IR64, Pusa and CA212 exhibited stronger inhibition of maximum photosynthetic rate (Pmax) and a decrease in Fv/Fm, which lead to the accumulation of AOS. It was found that the glutathione disulphide (GSSG) content in the glutathione pool and that of dehydoascorbate (DHA) in ascorbate pool of the leaves of these sensitive ones under chilling were induced to increase obviously. The correlation coefficient between the increase in GSSG, DHA and the decrease in chl content were 0.701 ** and –0.656 ** respectively. This indicated that the regeneration of reduced glutathione reduction of chl content, and the inhibition of photosynthetic activity. The changes in japonica Taipei 309 and Wuyujing were small. And the changes in indica hybrid were lying between the above mentioned types. Particularly, the ratio of ASA/DHA and GSH/GSSG showed similar changes as those in chl content. The correlation coefficient among chl content and ASA/DHA, GSH/GSS were significant.

It is common to find a build up of non structural carbohydrates in many parts of a plant under low temperature conditions, a response indicating that growth is more sensitive to low temperature than photosynthesis.

In response to extremes of temperature, plants can produce various proteins that protect them from damaging effects and falling rates of enzymes catalysis at low temperature so also the photorespiration. As temperature fall, production of anti freeze proteins rise. Cell membranes are also affected by changes in temperature and can cause membrane to lose its fluid properties and become a gel in cold conditions. This can affect the movement of molecules across the membrane. In cold conditions more unsaturated fatty acids are placed in the membrane. Various phenotypic symptoms in response to chilling stress include reduced leaf expansion, wilting, chlorosis and may lead to necrosis. Low temperature affects membrane and depends on the proportion of unsaturated fatty acids in it to the tolerance capacity so the conversion from fluidity to semi fluidity is less which leads to solute leakage (Mahajan and Tuteja 2005).

Influence of low temperature (chilling $5°C$ for 19hr) on photosynthesis ($mgco_2 dm^{-2} hr^{-1}$) and transpiration ($g.dm^{-2} h^{-1}$) of rice plant (Tanaka and Yoshitomi 1973) is given hereunder.

Table 11 Influence of chilling and oxygen level on photosynthesis ($mgco_2 dm^{-2} h^{-1}$) and transpiration ($gdm^{-2} h^{-1}$).

| | Before Chilling | | After chilling | | | | | |
| | | | 3 hr | | 4.5 hr | | 21 hr | |
Variety	$O_2\%$	P	T	P	T	P	T	P	T
Nankai 23	21	13.4	3.2	3.6	1.7	1.0	1.4	11.3	2.2
	3	17.9	3.0	0.9	1.5	-	-	17.8	2.4
IR8	21	14.5	3.3	1.2	1.6	0.7	1.4	3.7	2.1
	3	18.3	3.0	0.2	1.4	-	-	7.3	2.2

In Hunan Province in China, where double cropping of indica rice is common, low temperature damage on seedlings during the nursery period becomes a main problem in the first cropping. Effect of low temperatures (8, 11 and 14°C) on the root growth and the root secretion rate of the seedlings during raising stage of indica (Zhe9248, Xiang 24) Japonica I (Koshikikari) and japonica II (Taushinwase, Hayayuki) are characterized by their low temperature tolerances, suggests that root growth and the secretion rate of indica and japonica I were always lower to those of japonica II at each low temperature. In particular, it was noted that secretion rates of indica and japonica I have decreased promptly at treatment under 11°C, but the secretion rate of japonica II was relatively high even at 8°C. The recoveries of the root growth and secretion rate of exudates of japonica II were higher than those of indica and japonica I. The degrees of recovery in both traits revealed their close relationship with the quantity of secretion rate of exudates under preceding low-temperature condition. Based on these the low-secretion rate under low temperature may be responsible for the interior recovery of the root growth and the secretion rate of the exudates of indica and japonica I in comparision to japonica II. These suggest that the root membrane permeability is disrupted under low temperature specially in indica, japonica I than japonica II (Table 12).

Table 12 Means of percentage of saturated and unsaturated fatty acids of the three classes of cold reaction obtained after evaluation of 44 rice genotypes at the vegetative stage under low temperature treatments.

Cold reaction	Saturated fatty acids		Unsaturated fatty acids	
	28°C	10°C	28°C	10°C
Tolerant	28.7	28.6	72.0	72.6
Intermediate	29.1	29.2	72.0	70.5
Sensitive	29.3	31.8	71.1	68.7

(after Cruz et al. 2010)

The critical air temperature for rooting is reported as 14–15.5°C for seedlings in wetland seedbeds, 14–14.5°C for seedlings from semi-irrigated seed bed and 13–13.5°C for seedling from a dry land seed bed (Yatsuyanagi 1960).

When the seedlings of two rice cultivars, IR8 (low-temperature sensitive) and Somewake (low-temperature tolerant) were exposed to a low temperature of 15°C, the normal increase in the chlorophyll content of the developing 4th leaf blade ceased completely while increases in protein content continued at a low rate in both cultivars. Analysis of soluble and insoluble proteins in the 4th leaf blade of IR8 by SDS-polyacrylamide gel electrophoresis revealed that synthesis of RuBP carboxylase and several thylakoid proteins responsible for photosynthetic electron transport and photophosphorylation was greatly inhibited at low temperature. It was also found that increases in the activities of some enzymes of the Calvin Cycle, such as RuBP carboxylase, fructose bisphosphatase and NADP-glyceraldelye -3-P dehydrogenase as well as of catalase were specifically inhibited during growth at the low temperature.

This results suggest that the synthesis of intracellular components, in particular of key proteins required for photosynthesis, is specially susceptible to low temperature stress during development of rice leaves (Maruyama et al. 1990).

When seedlings of the rice cultivar K-Sen 4 were exposed, at the germination and the leaf stages, to 5°C for 7 days, they withered after incubation at 25°C. In contrast, the cultivar Dunghan Sali showed chilling tolerance and successful growth after rewarming (Saruyama and Tanida 1995). They also tried to correlate the difference in cold sensitivity with superoxide dismutase (SOD), catalase (CAT), ascorbate peroxidase (APX), and glutathione reductase (GR) (activated oxygen-scavenging system). NADP-dependent isocitrate dehydrogenase (NADP-ICDH) (TCA cycle), and glucose-6-phosphate dehydrogenase (G6P-DH) (Pentose phosphate cycle) activity. CAT activities in both cultivars were drastically decreased, by the chilling. In Dunghan Shali, these activities were recorded and stimulated by rewarming. However, in K-Sen4, the rewarming decreased the activities in the embryo and root at the leaf stage. For APX at the germination stage, chilling resistance was found with both cultivars, but the rewarming enhanced the activities effectively in Dunghan Shali while not as effectively in K-Sen 4. APX activities at the leaf stage in both cultivars were not affected very much by the chilling. However, only 10% of the initial activity was detected in K-Sen 4 root after rewarming. The other enzymes, SOD, GR, ICDH, G6P-DH, displayed no significant differences in cold sensitivities between the two cultivars. It is concluded that the tolerance of rice cultivars to chilling injury is closely linked to the cold stability of CAT and APX.

It is also noted, that the analysis of the isozyme profile and activity of superoxide dismutase (SOD), catalase (CAT), ascorbate peroxidase (APX) and glutathione reductase (GR) suggests that significant induction of expression and activity of antioxidative enzymes CAT and APX in the leaves and SOD, CAT, APX are most important for cold acclimatization and chilling tolerance. Increased activity of antioxidants in roots is more important for cold tolerance than increased activity in shoots (Kuk et al. 2003).

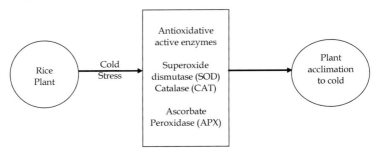

At low temperature glutamine synthatase and isocitrate dehydrogenase activities were increased by 80 and 75% respectively thus a positive parallel correlation was observed between the two enzymes. Native PAGE analysis, together with activity staining and Western blot assays, showed that both GSrb activity and GSrb protein level were enhanced under low temperature. Meanwhile, NADHGDH and NADGDH activity were both reduced, though to different extents. At low temperature, ammonium absorption was stimulated by the elevation of GS activity. Proline content was increased twofold by low temperature and this accumulation was in good agreement with the induction of GS activity. Similar observations were made in the rice roots fed with sucrose, indicating that at least partially low temperature has the same effect as carbon compounds on the modulation of nitrogen metabolism (Lu et al. 2005).

Polyamines are ubiquitous low molecular weight aliphatic that are involved in regulation of plant growth and development (Martin–Tanguy 2001). Because of their polycationic nature at physiological pH. Polyamines are able to interact with proteins, nucleic acids, membrane phospholipids and cell wall constituents, thereby activating or stabilizing these molecules. The most commonly found polyamines in higher plants, the diamine putrescine (Put), the triamine spermidine (Spd) and the tetramine supermidine (Spm) may be present in the free, soluble, conjugated and insoluble bound forms (Fig. 21).

Cold stress stimulated the phosphorylation of a 60K da protein in the cold sensitive rice variety IR 36, in an *in vitro* study. However, in the cold tolerant rice variety, Kitaibuki, this protein had already been phosphorylated (Komatsu and Kato 1997).

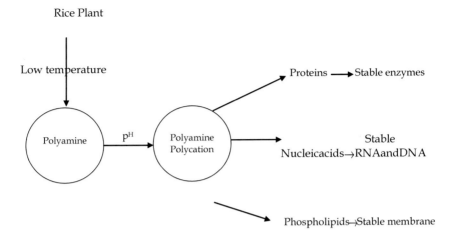

Fig. 21 Protective mechanism under cold stress.

2-week old rice seedlings under 13/15°C for 14 days showed that root and shoot dry weights were significantly less from the control condition (20/22°C), night and day temperature. Results for stomatal conductance and the SPAD value from the control condition were significantly more than those from the stress condition. The leaf glucose content from the stress condition was more than that from the control indicating restriction in GA3 synthesis. ABA increase led to a decrease of leaf glucose and stomatal conductance. In addition, low temperature stress caused a decrease of photosynthetic capacity. High concentrations of glucose, increased ABA biosynthesis. Glucose induced ABA biosynthesis and mediated some signaling mechanisms for stomatal reactions. An increase of leaf glucose from the stress condition caused IAA protection against oxidative reaction in the stress tolerant genotypes.

ABA has a key role among plant hormones in controlling damage from low temperature stress. Glucose has a direct and extensive role in the regulation of ABA biosynthesis

(Leon and Sheen 2003). A high glucose concentration leads to high concentrations of ABA within cells (Arenas–Huertero et al. 2000) and provides increasing gene transcription of ABA biosynthesis (Cheng et al. 2002). In addition, sucrose, like glucose is also necessary for the regulation of ABA biosynthesis. Cytokinins are often considered abscisic acid antagonists and auxins antagonists/synergists in various processes in plants (Table 13).

Low temperature stress caused significant increase in ascorbate (total, oxide and reduced) and a tocopherol levels in both roots and leaves. Both roots and leaves of cold tolerant variety IRCTN 33, ascorbate oxide was significantly higher than other varieties. Leaf hydrogen peroxide in sensitive variety, Hoveizeh had significantly increased, which had negative correlation with the amount of chlorophylls. Significant increase of malondialdehyde in Hoveizeh leaves showed severe membrane phospholipid peroxidation. In the leaves the least amount of H_2O_2 was observed in cold tolerant genotype IRCTN34. The roots ascorbate oxide form in IRCTN34 was significantly less than other genotypes.

However, Hoveizeh had weak antioxidant mechanism among studied genotypes, while that of low temperature tolerant genotypes were highly efficient. Therefore, it is possible to recommend them as tolerant parent for rice low temperature breeding (Hassibi 2010).

Both biotic and abiotic stresses can result in oxidative stress through the formation of free radicals, which are highly destructive to lipids, nucleic acids, and proteins (Mittler 2002). Water stress, which is also produced as a secondary stress by chilling (Levitt 1980).

Table 13 Correlation of different characters of rice genotypes in control and stress conditions

	Shoot dry weight	SPAD Value	Stomatal Conductance	Leaf IAA	Leaf ABA	Leaf glucose	Root dry weight	Root IAA	Root ABA	Root glucose
Shoot dry weight	1									
SPAD Value	-0.54	1								
Stomatal conductance	-0.55	0.58**	1							
Leaf IAA	-0.59*	0.40	0.20	1						
Leaf ABA	-0.16	0.40	-0.52*	0.59**	1					
Leaf glucose	0.66**	-0.52**	-0.70**	0.09	-0.13	1				
Root dry weight	0.82**	0.63**	-0.28	-0.43	-0.23	0.59*	1			
Root IAA	-0.23	0.12	0.22	-0.06	-0.19	-0.50*	-0.20	1		
Root ABA	0.11	0.23	0.24	-0.45*	-0.14	-0.59**	0.19	0.05	1	
Root Sugar	-0.15	0.18	0.00	-0.25	-0.06	-0.28	-0.19	0.190	0.78**	1

Membrane	H_2O_2	malondialdehyde
Phospholipid	peroxidase	

	Tolerant (IRCTN 33)	Sensitive (Hoveizeh)
Low temperature	High ascorbate	low ascorbate
	High Chlorophyll	High Chlorophyll
	Low malondialdehyde	High malondialdehyde
	Less hydrogen peroxide	High hydrogen peroxide
	Strong antioxidant mechanism	Weak antioxidant mechanism

Temperature acclimation results from a complex process involving a number of physiological and biochemical changes, including changes in membrane structure and function, tissue water content, global gene expression, protein, lipid and primary and secondary metabolite composition (Shinozaki and Dennis 2003). Recent advances in genome sequencing and global gene expression analysis techniques have further established the multigenic quality of environmental stress responses and the complex nature of temperature acclimatization (Kreps et al. 2002).

Cold shock influenced metabolism profoundly, the steady state pool sizes of 311 metabolites or mass spectral tags were altered in response to cold shock. Increases in the pool sizes of amino acids derived from pyruvate and oxaloacetate, polyamine precursors, and compatible solutes were observed during cold shock, known as signaling molecules and protectants (Kaplan et al. 2004) (Fig. 22).

Signal perception is the first step of plant response to environmental stress. A stress sensor can detect environmental variables and transmit the initial stress signals in cellular

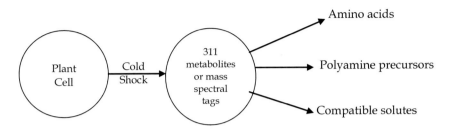

Fig. 22 Cold influence on metabolites.

targets specifically. Some of the two-component histidine kinases function as stress sensors in bacteria and yeast. Each environmental stimulus provides plant cells with specific yet related informations.

Given a large number of potential stimuli, it is possible that plants may monitor the unique attribute of stress signals through different kinds of sensors. Once an extra cellular stimulus is perceived, second messenger molecules, etc., e.g., Ca^{2+}, inositol phosphates and reactive oxygen species (ROS) are immediately generated. Second messengers subsequently activate a downstream signal cascade that phosphorylates transcription factors that regulate the expression of a set of genes or proteins involved in stress adaptation. Phosphorylation by protein kinases is the most common and important regulatory mechanism in signal transduction (Gao et al. 2008).

Stimuli	Low temperature (12°C)
	↓
Second Messenger	?
	↓
MAPKK	?
	↓
MAPKK	OsMEK 1
	↓
MAPK	OsMAP 1

According to Mahajan and Tuteja (2005) the mechanism of receptor sensor mediated signaling for transcription may be as follows:

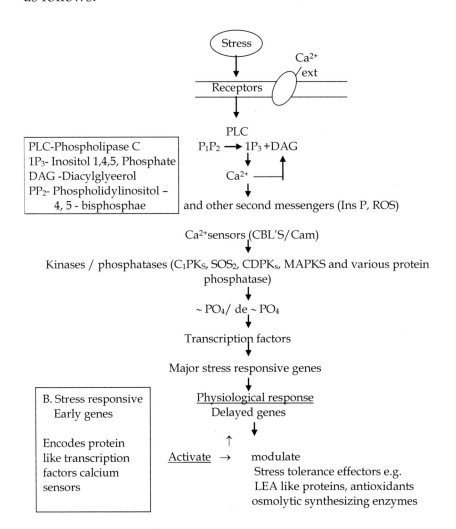

4

REPRODUCTIVE DEVELOPMENT

Frequently occurring low temperature causes more than 50% yield loss. Diseases such as blast and sheath brown rot adds to the damage especially, when it attacks at the plants early vegetative stage. In China, the recorded yield loss due to low temperature is 3–5 million tons. In 1980, Korea lost an average yield of 3.9 tons per hectare.

Cold tolerance is a complex trait controlled by many genes. IRRI scientists have identified three regions of the rice genome that have a direct link to cold tolerance at the plant's reproductive stage. Cold stress at critical times of reproduction hinders the formation of fertile pollen that is crucial for fertilization and consequently the rice plant may fail to produce grains.

Panicle initiation is the start of the reproductive phase of rice development. It is when the actual panicle or head begins to form in the base of the shoots or stems, just above the soil surface. The formation of the panicle marks the end of tillering or vegetative phase and the beginning of the reproductive phase. Panicle initiation is defined as when three out of ten main shoots have a panicle 1 to 2 mm long, above the airpace or internode. The early changes associated with the start of panicle initiation are microscopic in size and impossible to detect with the naked eye. Panicle initiation depends on temperature, variety, nitrogen fertility, plant population and water depth during tillering.

The reproductive phase, i.e., panicle initiation to flowering is about 35 days and ripening phase is about 30 days on an average.

Panicle formation is the change from vegetative apical meristem to a reproductive apical meristem. Panicle exsertion is the exsertion of the panicle above the flag leaf sheath after anthesis. It is generally accepted that cold tolerance of rice at one stage is different from another stage. Panicle formation is interrupted by the low temperature (Okabe and Toriyama 1972). Days to complete heading varies with the variety, the range being 5–15 days. By comparison, vegetative shoot apical meristem is very small, usually completely hidden within leaf sheath. The panicle exsertion can be classified as well exserted, moderately well exserted, just exserted, partly exserted and enclosed.

In aman rice cultivation deviation from the optimum planting time may cause incomplete and irregular panicle exsertion, increased spikelet sterility (Magor 1984). Under temperate condition or in hill altitude similar results are noted in many varieties.

Rice is very sensitive to cool temperatures during the reproductive period (Yashida 1981). Although the main symptom of damage from cold is the high spikelet sterility (Jacobs and Pearson 1994), incomplete panicle exsertion has been cited as a symptom as a cold injury in many countries (Chung 1979, Hamdani 1979, Alvarado and Grau 1991). Cold tolerance at reproductive period has been associated with the degree of panicle exsertion, which could be used as a selection criterion. In fact, this trait was suggested as an indicative of genotype adaptability to cool temperatures (Takahashi 1984).

During reproductive phase two types of spores, microspores and megaspores are produced that give rise to male gametophytes and female gametophytes, respectively. The fusion of egg and sperm gives rise to the zygote, which is the beginning of diploid sporophyte generation thereby completing the lifecycle. The female gametophyte, also referred to as the embryo sac or mega gametophyte, develops within the ovules, which is formed within the carpel's ovary, consists of seven cells and four different cell types; three anti podal cells, two synergid cells, one egg cell and one central cell.

At panicle initiation stage microsporogenesis was determined by the distance between the ligule of the flag leaf and that of the penultimate leaf (Yoshida 1981) considering an interval of –3 (Flag leaf ligule below the penultimate leaf ligule) to +10cm (Flag leaf ligule above penultimate leaf ligule) as indicative of this stage. Maximum microsporogenesis may vary, however within their range, depending on the genotypes. In some genotypes it may take place at –3 cm, in others at zero and in others at +8 cm, indicating that genotype, reproductive stage and duration of cold exposure differed for the percentage of reduction in panicle exsertion and in spikelet sterility under 17°C.

In all durations of cold exposure, reduction in panicle exsertion was generally larger at anthesis (Anthesis was considered as the beginning of panicle exposure) than at microsporogenesis. There is no clear distinction between cold tolerant and sensitive genotypes in regard to reduction in panicle exsertion in any of the stages and duration of cold exposure, as this was the most sensitive stage to cold when cold tolerance was measured as the reduction of spikelet fertility. In all durations of cold exposure and stages, indica genotypes were more sensitive than japonica in considering percentage reduction in spikelet fertility (Cruz et al. 2006).

Cold treatment at the young microspore stage results in a dramatic depletion of starch accumulation in the pollen grains at anthesis. Starch is essential in pollen grains, as storage carbohydrate for energy supply during pollen germination and fertilization.

Measurements of sucrose content showed that sucrose levels increased in anthers that were cold–treated at the young microspore stage. Consistent with these results, it was found that the enzyme activity levels of cell wall invertase, but not soluble invertase decreases upon cold treatment at the young microspore stage only.

These results further suggest that cold induces a reduction of cell wall invertase activity, which then leads to a blockage of sucrose supply to developing pollen grains. This blockage is explained by the absence of starch synthesis in the pollen grains at maturity (Fig. 23).

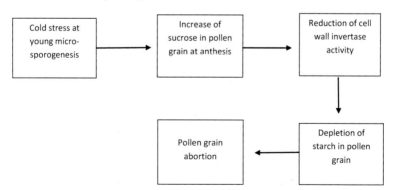

Fig. 23 Effects of low temperature on pollen grain.

Many self fertilizing crops are particularly sensitive to abiotic stress at the reproductive stage. In rice and wheat, for example, abiotic stress during meiosis and young microspore stage indicates the developmental program appears to be compromised. Tapetal hypertrophy can occur as a consequence of cold and drought stress, and programmed cell death (PCD) is delayed or inhibited. Since the correct timing of tapetal PCD is essential for pollen reproduction, substantial losses in grain yield occur. In wheat and rice, a decrease in tapetal cell wall invertase levels is correlated with pollen abortion and results in the amount of hexose sugars reaching the tapetum, and subsequently the developing microspores, being severely reduced. ABA and gibberellins levels may be modified by cold and drought, influencing levels of cell wall invertase(s) and the tapetal developmental program, respectively. Many genes regulating tapetal and microspore development have been identified in Arabidopsis thaliana (L.). Heynh and rice and the specific effects of abiotic stresses on program and pathways can now begin to be assessed (Parish et al. 2012).

Chilling during male gametophyte development in rice inhibits development of microspores, causing male sterility. Changes in cellular ultrastructure that have been exposed to mid chilling include microspores with poor pollen wall formation, abnormal vacuolation and hypertrophy of the tapetum and unusual starch accumulation in the plastids of the endothecium in the post meiotic anthers. Anthers observed during tetrad release also have callose(1,3-β-glucan) wall abnormalities (Fig. 24).

Expression of rice anther specific monorepressed genes(OsMST8) is greatly affected by chilling treatment. Perturbed carbohydrate metabolism, which is particularly triggered by repressed genes OsINV4 and OsMST8 during chilling, causes unusual starch storage in the endothecium and

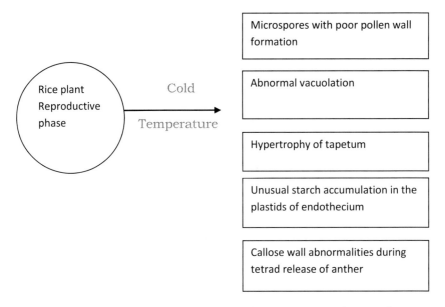

Fig. 24 Ultrastructure changes during cold stresses at reproductive phase.

this also contributes to other symptoms such as vacuolation and poor microspore wall formation. Premature callose breakdown apparently restricts the basic framework of the future pollen wall. Vacuolation and hypertrophy are also symptoms of osmotic imbalance triggered by reabsorption of callose breakdown products due to absence of OsMST8 activity (Mamun et al. 2009).

Low temperature (12°C), for 4 days during microsporogenesis and anthesis produced considerable sterility. Sensitivity was greatest when the florets from the midsection of the panicle were passing through the early microspore phase of pollen development. The amount of low temperature induced sterility was increased by high nitrogen supply (Heenan 1984).

Microspores are symplastically isolated in the locular space of the anthers, and thus an unloading pathway of assimilates via the apoplasmic space is mandatory for proper development of pollen. Antisense repression of the anther specific cell wall invertase or interference with invertase activity by expressing a proteinacious inhibitor under the control of anther-specific invertase promoter results in a block during early stage of pollen development, thus causing male sterility without having any pleiotropic effects. Restoration of fertility was successfully achieved by substituting the down-regulated endogenous plant invertase activity by a yeast invertase fused with the N-terminal portion of potato derived vacuolar protein proteinase II (Pi II-Scsuc2), under control of the orthologous anther specific invertase promoter Nin88 from tobacco (Engelke et al. 2010).

In rice, a female archesporial cell elongates longitudinally and directly differentiates the megasporocyte or megaspore mother cell (MMC). The MMC then undergoes meiosis, resulting in the formation of four haploid spores. A chalazal spore becomes a functional megaspore, where as three spores toward the micropyle undergo programmed cell death.

In the majority of the flowering plants, the dynamic phase transition of sporophyte to gametophyte occurs in the anther and ovule. Sporogenosis is characterized by the differentiation of hypodermal cells in anther and ovule primordia, termed primordial germ cells or archesporial cells, into microsporocytes and megasporocytes and megaspores, in which a differential pattern of spore formation between male and female organs called heterospory, usually observed in flowering plants, including rice. Low temperatures during this stage disrupt proper pollen developments, leading to a shortage of sound pollen at the flowering stage (Satake 1976, Farrell et al. 2006).

CSIRO is determining the molecular basis of pollen sterility in rice. Cold repress sugar transport to the pollen grains, causing abortion of pollen development. Two important genes that are involved in sugar transport have been identified and both genes are repressed by cold (CSIRO 2009). CSIRO is locating and characterizing useful rice genes and addressing rice tolerance to cold.

Pollen mother cells (PMC) under cold stress appeared normal before meiosis, but during meiosis, normal callose deposition was disrupted.

Consequently, the PMCs began to degenerate at the early meiosis stage, eventually resulting in complete pollen collapse. In addition, the degeneration of the tapetum and middle layer was inhibited. These results demonstrate that rice Ugp1 is required for callose deposition during PMC meiosis and bridges the apoplastic unloading pathway and pollen development (Fig. 25). UDP-glucose pyrophosphorylase (UDPase) catalyzes the reversible production of glucose-1-phosphate and UTP to UDP-glucose and pyrophosphate. The rice genome contains two homologous UGPase genes, Ugp1 and Ugp2 (Chen et al. 2007).

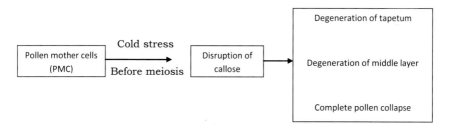

Fig. 25 Effects of cold stress before meiosis on pollen mother cells.

Low temperature may also result in non-receptive stigma, fertilization failure or post fertilization development. These important aspects of flowering behavior in rice were studied by Nizigiyimana (1990) who also reported that low temperature sterility could be due to degeneration of embryo sac, malformation of ovaries and deformation of young sporocytes during meiosis (Fig. 26).

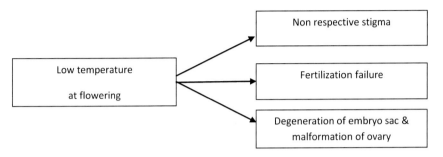

Fig. 26 Influence of low temperature at flowering.

Since, the florets of rice are adichogamous, most of the florets are self pollinated at the time of floret opening. Synchrony between floret opening and anther dehiscence may contribute to the high rate of self-pollination. However, the synchronization is not always perfect and some rice florets hybridize naturally. The rate of natural hybridization varies among varieties of rice suggesting that there is a difference in the reliability of self-pollination among varieties.

Rice pollination is susceptible to meteorological factors, such as temperature. High temperatures at the time of flowering inhibit the swelling of pollen grains, whereas low temperature at the booting stage impede pollen growth. The driving force for anther dehiscence is the swelling of pollen at the time of floret opening, temperature stress is considered to reduce the percentage of anthers dehiscing at the time of flowering. Thus high (>35°C) and low (<20°C) temperatures result in poor pollination and loss of yield (Table 14).

Table 14 Percentage of florets with more than 80 and 40 pollen grains deposited on the stigma and number of pollen grains deposited on the stigma.

Cultivar	Florets with >80% pollen Grains on the stigma (%)	Florets with>40% pollen Grains on the stigma(%)	Number of pollen grains deposited on the stigma Mean C.V.
Mubouai koku, Indigenous	26.4 ± 21.0	53.3 ± 21.7	59.3 ± 27.7 84.1 ± 5.4
Somewake, Indigenous	77.3 ± 15.3	97.3 ± 6.0	137.0 ± 23.5 53.9 ± 15.9
Homura 3, Breeding station	97.4 ± 3.5	100.0 ± 0.0	204.1 ± 60.2 44.4 ± 7.3
Magatama, Breeding station	87.8 ± 9.8	100.0 ± 0.0	194.6 ± 91.9 48.6 ± 3.5

(After Matsui and Omasa 2002)

Morphologically, large anthers and longer stigmas (Suzuki 1982) contribute to increased tolerance to cold stress during the booting stage.

Spikelet opening triggers rapid pollen swelling, leading to anther dehiscence and pollen shedding from the anther's apical and basal pores (Matsui et al. 1999). Increased basal pore length in a dehisced anther was found to contribute significantly to successful pollination (Matsui and Kagata 2003), probably because of its proximity to the stigmatic surface. Longer stigmas may also be important for the same reason. After pollination, it takes about 30 minutes for the pollen tube to reach the embryo sac. Genotypic differences in pollen number and germinating pollen on stigma and spikelet fertility was affected by cold stress.

Low temperatures affect the development of pollen grains, decreasing the number of engorged ones, preventing fertilization and causing spikelet sterility, which in turns lowers grain yields (Lee 2001). Gunawardhana et al. (2003)

found that low temperature during very early and peak microspore development caused a severe reduction of total pollen production.

The development of male reproductive organs in rice is very sensitive to various environmental stresses. For example, exposing plants to low temperatures during the heading stage leads to a reduction in gram yield. It was found that rice grown under normal conditions and also at three different temperatures 16, 18 and 20°C significantly decreased pollen viability and grain production. Cytological observations of the anther showed that the tapeum was the most sensitive to low temperature stress, resulting in male sterility due to functional loss of the tissue. Electron microscopy suggested that this abnormality was restricted primarily to the endosplasmic reticulum (ER), a highly vulnerable sub-cellular organelle, showed two typical morphological aberrations, one can use it as a pattern of arrangement, the other in the formation of ER bodies, the most severe abnormalities were noted in tapetal cells exposed to 16°C (Fig. 27).

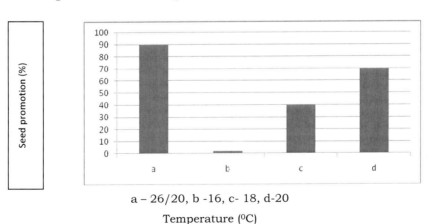

a – 26/20, b -16, c- 18, d-20

Temperature (⁰C)

Fig. 27 Some seed formation changes by cold stress on rice at heading.

Low temperature during flowering may also cause spikelet sterility, however its effects are not as pronounced as the ones caused by cold during pollen formation. Based on data reported by Alvarado (1999), a decrease in temperature of one degree, below the threshold air temperature of 20°C, increases spikelet sterility by 6%.

After low temperature (16°C) treatment, the net photosynthetic rate, chlorophyll content and FBPase activity tended to decrease and ratio of chlorophyll a/b, stomatal resistance, sucrose and starch content tended to increase in rice flag leaves at heading stage. However, the Pi content and water potential did not show pronounced change. The flag leaf chloroplasts contained large starch grains after three days at low temperature (Jie et al. 1981) (Fig. 28).

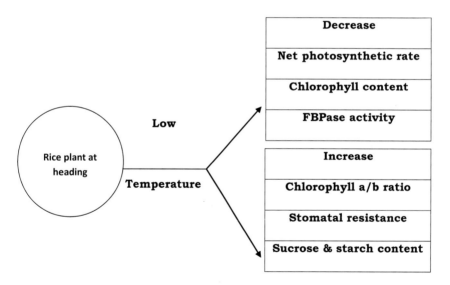

Fig. 28 Some physiological changes by cold stress on rice at heading.

In rice, stomatal aperture as well as conductance is strongly correlated with leaf photosynthesis where as photosynthesis by rice leaves is also influenced by other factors, such as leaf nitrogen and content of the enzyme ribulose-1, 5- bisphosphate carboxylase/oxygenase; stomatal conductance is codominantly correlated with the rate of leaf photosynthesis (Makino et al. 1987).

Temperature range of 10 to 15°C has been reported to significantly reduce chlorophyll content, and below 10°C, the chlorophyll content reduction is more pronounced due to the fact that membrane bound chlorophyll is destroyed by the free radicals of oxygen despite the protective action of carotenoids (Smillie et al. 1987).

Photosynthesis, a key determinant of the rate of plant growth, is influenced by both CO_2 concentration and is also related parabolically to leaf temperature. These responses are described mechanistically by the biochemical model of photosynthesis. The model has two major parameters: the potential rate of electron transport (Jmax) and the maximum rate of RuBP (ribulose 1,5 bisphosphate) carboxylation (Vc max). Photosynthesis often shows down regulation under a long term increase in CO_2 concentration. In many species, a long term increase in temperature leads to all increase in the optimal temperature of maximal photosynthetic rate. There was both an increase in the absolute value of the ligh saturated photosynthetic rate at growth CO_2 (P growth) and an increase in Topt for P growth caused by elevated CO_2 in FACE conditions. Seasonal decrease in P growth was associated with a decrease in nitrogen content per unit leaf area (N area) and they in the maximum rate of electron transport (Jmax) and the maximum rate of RuBP Carboxylation (Vc max). At ambient CO_2, Topt.

increased with increasing growth temperature due mainly to increasing activation energy of Vc max.

Canopy temperature and incident photosynthesis active radiation were provided in mature plants of five varieties of rice exposed to a 3 day chilling treatment (21°C day/10°C night) 14 days after anthesis. Canopy photosynthesis was measured before, during and after the chilling period. A Japonica like cultivar Hungarian-1 from Central Europe, and an indica like cultivar Lemonet from Texas were markedly less sensitive to a chilling dependent reduction in photosynthesis than were cultivars Guichao-2, Er. Bai Ai, and IR8 from South east China and the philippines. The extent of chilling dependent reduction in photosynthesis was greater if roots of plants were allowed to be at air temperature in night. This led to a 5–7°C differential between root and air temperature in the morning and associated with a midday depression of photosynthesis in all varieties. If root temperature was kept at 20°C during the day and night and air temperature controlled at 21°C day and 10°C night, the effects of chilling on photosynthesis were much less pronounced.

Changes in canopy photosynthesis following treatments were corelated with change, in leaf level perameters. The level of sugars and starch levels was found to remain high during the first cool night and to increase further during the next day. In cultivars which showed pronounced inhibition of canopy photosynthesis throughout the chilling treatment (Guichao2, IR8) soluble sugars remained at high levels until temperatures were returned to control values. In the less sensitive cultivar Hungarian-1 soluble sugars declined throughout the chilling treatment. It is suggested that inhibition of photosynthesis may be associated with sugar retention in leaves of rice at low temperatures, and imply differences between cultivars in the response. Hungarian-1 showed least impairment in photosynthesis with greater reduction in quantum yield under chilling treatment.

Rice plants lose significant amounts of volatile NH_3 from their leaves; an increase in O_2 increased ammonia diffusion rate in the two cultivars, accompanied by a decrease in photosynthesis due to enhanced photo respiration, but did not greatly influence transpiration rate and stomatal conductance. There were significant positive correlations between ammonia emission rate and photorespiration in both cultivars.

Increasing light intensity increased AER, PG, Tr and Gs in both cultivars, where as increasing leaf temperature increased AER and Tr but slightly decreased PG and Gs. Thus, photorespiration is strongly involved in NH_3 emission by rice leaves and with differences in GS activity. It is also suggestive that NH_3 emission in rice leaves is not directly controlled by transpiration and stomatal conductance (Fig. 29).

Ammonia emission from rice leaves is in reaction to photorespiration and genotypic differences in glutamine synthetase activity (Kumagai et al. 2011).

Temperature acclimation for photosynthesis has been related to leaf nitrogen economy, since more than half of the leaf nitrogen is in photosynthetic apparatus and thus the photosynthetic capacity is strongly related to the leaf nitrogen content. Leaf nitrogen content on a leaf area basis is greater in leaves grown at lower temperatures. This is considered as a compensatory response to low temperature, which decreases enzyme activity. As respiration rates are also related to leaf nitrogen content, a similar response has also been observed for the temperature acclimatization of respiration.

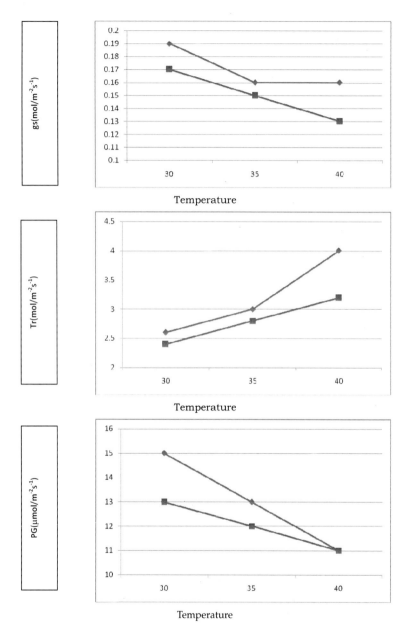

Fig. 29 Changes in stomatal conductance, transpiration rate and photosynthesis in leaves of rice genotypes at heading.

Color image of this figure appears in the color plate section at the end of the book.

For photosynthesis, not only leaf nitrogen content but also nitrogen use within a leaf is suggested to be related to the temperature acclimatization. Changes in nitrogen partitioning among photosynthetic components can be a factor responsible for changes in the temperature dependence of the photosynthetic rate, and thus may affect the photosynthetic rate at the growth temperature. It has also been reported that the temperature dependence of Rubisco kinetics changes with growth temperatures.

Temperature dependence of the respiration rate is considered to be determined by the maximum activity of respiratory enzymes, availability of substrates and/or demand for respiratory energy. It has been reported that temperature acclimatization of respiration involves increases in respiratory capacity by increasing the capacity per mitochondrion or increasing the number of mitochondria, the mitochondrial density and the density of cristae within mitochondria. Thus, temperature acclimatization of respiration could be linked to changes in the enzyme capacity (Yamori et al. 2009).

Much of the energy and carbon skeletons necessary for biosynthesis and cellular maintenance are produced by plant respiration. Under some conditions (e.g., excess irradiance) respiration (R) may also help minimize the formation of potentially damaging reactive oxygen species (ROS) through oxidation of excess cellular redox equivalents (Maxwell et al. 2009). R is also crucial for (1) the production of ascorbate (Millar et al. 2003), a necessary component of the protective xanthophyll and glutathione cycles, (2) the maintenance of photosynthetic activity, largely because of the energy demands of sucrose synthesis (Kromer 1995), and (3) regulating pathogen defence processes. R also plays an important role in determining the carbon budget of individual plants and the concentration of

CO_2 in the atmosphere. Between 30 and 80% of the CO_2 taken up by photosynthesis (P) each day is subsequently respired.

One of the most important environmental parameters affecting rates of plant R is temperature. It is often assumed that the relationship between plant R and temperature is exponential with a constant Q_{10} (about 2). The importance of understanding variability in the Q_{10} is highlighted by predicted changes in the amplitude of diurnal temperatures experienced by plants (Fig. 30).

The reduction in chlorophyll content has been implicated to metabolic blocks in the porphyrin pathway that leads to chlorophyll synthesis. Low temperature stress also affects enzymes that carry out photosynthetic processes in the plant. Because enzymes that are routinely involved in photosynthesis have little energy, the rate of photosynthesis is slowed down under low temperature. Satake and Hayase (1970) observed higher cytological and histological disorders in the anthers of cold stressed rice plants as compared to non-stressed plants. Cold temperature stress at reproductive stage has also been reported to cause flower abscission, pollen sterility, pollen tube distortion, ovule abortion and poor seed set. Structural and functionalities in the reproductive organs of cold-stressed plants and failed fertilization or premature abortion of the embryo have been observed.

The flower of rice diverged from those of model edicot species such as Arabidopsis, etc. and is thus of great interest in developmental and evolutionary biology. Specific to grass species including rice, are the structural units of the inflorescence called the spikelet and floret, which comprise grass-specific peripheral organs and conserved sexual organs. Recent advances in molecular genetic studies have an

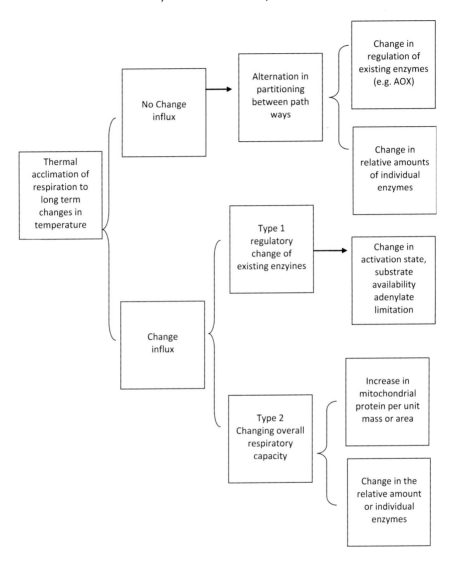

(after Atkin et al. 2005)

Fig. 30 Respiration change mechanisms.

understanding of the functions of rapidly increasing number of genes involved in rice flower development. The genetic framework of rice flower development is in part similar to that of model edicots. However, rice also probably recruits specific genetic mechanisms, which probably contribute to the establishment of specific floral architecture of rice (Yoshida and Nagato 2011).

The initiation of panicle primordium at the tip of the growing shoot marks the start of the reproductive phase. The panicle primordium becomes visible to the naked eye about 10 days after initiation. At this stage, 3 leaves will still emerge before panicle finally emerges.

In short duration varieties the panicle becomes visible as white feathery cone 1.0–1.5 mm long. It occurs first in the main culm and then in tillers where it emerges in uneven pattern.

The young panicle that emerges inside the bottom of the last node is first a little feathery cone shaped organ 1–1.5 mm, which is visible only if the stem is dissected. In fact, the cone becomes visible only about 10 days after it is formed. At this stage the number of spikelets in the panicle is already determined. In short duration varieties, maximum tillering, internode elongation, and panicle initiation occur almost simultaneously. These stages occur in the above mentioned order in medium to long duration varieties. Timing of panicle initiation in rice is influenced by many factors, among which some constants are inherent to variety, temperature and photoperiod (Fig. 31).

Root growth measured as root length, is rapid and linear during vegetative development of the rice plant. Maximum root length is observed by panicle initiation or booting and is maintained at nearly constant level until heading following heading, root length declines until milk stage where it may

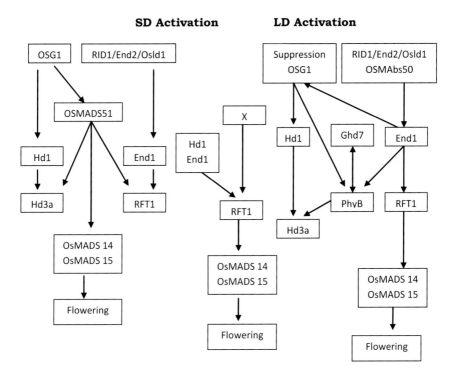

Fig. 31 Variety, temperature and photoperiod in flowering.

remain at a reduced level or may increase by physiological maturity. However, the relationship of root growth to plant development was poorly understood.

Anthesis or flowering refers to the events between the opening and closing of the spikelet (floret) and lasts for 1 to 2½ hours. Flowing generally begins upon panicle expiration or on the following day and is consequently considered synonymous with heading. Anthesis generally occurs between 9 a.m. to 3 p.m., but varies at places. The steps involved are:

a) The tips of the lemna and palea (hulls) open.
b) The filaments elongate.

c) The anthers exert from the lemna and palea.
d) As the lemna and palea open further, the tips of the feathery stigma become visible.
e) The filaments elongate past the tips of lemna and palea.
f) The spikelet closes, leaving the anther outside. Anther dehiscence (pollen shed) usually occurs just prior to or at the time the lemna and palea open.

Pollen grains are visible for about five minutes after emerging from the anther, whereas the stigma may be fertilized for three to seven days. Rice is a self-pollinating plant, because it is usually pollinated before the lemna and palea open to release pollen into the air. Fertilization of the ovary by the pollen grain is generally completed within five to six hours after pollination; at that point, the overy becomes brown rice. The time interval for flowering of an entire panicle is normally four to seven days. A relationships between plant and soil factors that control the availability and uptake of mineral nutrients is that soil temperature affects nutrient uptake directly by altering root growth, morphology, and uptake kinetics. Indirect effects include altered rates of decomposition and ineral nutrient mineralization, mineral weathering, and nutrient transport processes (mass flow and diffusion) (Fig. 32).

It is noted that a flower-specific proline rich protein mainly present in rice flower and accumulated abundantly during the late stage of flower development which is associated with cold tolerance in rice, a cell wall protein, playing a crucial role in determining extracellular matrix structure of floral organs (Gothandam et al. 2010).

In general the mechanism of cellular response to induce cold stress in rice at the reproductive phase is similar to the mechanism described earlier and is mediated via ABA synthesis and the cascade of changes in metabolism to hinder the

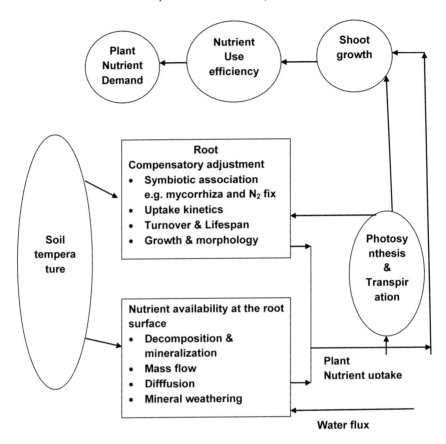

Fig. 32 Concept on nutrient uptake.

development process. In the process cytokinin content declined while ABA content increased during cold acclimatization, in spite of a huge cellular division and enlargement which is associated in the reproductive period of rice growth.

From the above mentioned findings it is evident that low temperature at early and late reproductive phase of plant growth of rice, ultimately, spikelet fertility is affected severely. Rice varieties respond differently the spikelet sterility at any of the two or both development stages of reproduction. It can also be noted that there are varieties, in which though low temperature does not affect the growth appreciably, spikelet fertility is seriously affected. On the other hand there are varieties which are also less damaged by low temperature. However, there are few varieties which are tolerant to low temperature at reproductive stage.

5

GRAIN DEVELOPMENT

Agronomically, heading is usually defined as the time when 50% of the panicles have exserted. Anthesis normally occurs between 0800 and 1300 hours in tropical environments. Fertilization is completed within 5–6 hours later. Only a very few spikelets have anthesis in the afternoon. When the temperature is low, however, anthesis may start late in the morning and last until late afternoon. Within the same panicle it takes 7–10 days for all the spikelets to complete anthesis; most spikelets complete anthesis within 5 days.

The length of ripening, largely affected by temperature, ranges from about 30 days in the tropics to 65 days in cool, temperate regions such as Hokkaido, Japan and New South Wales, Australia.

Upon fertilization, female gametophyte expressed genes control the initiation of seed development. During seed development, female gametophyte expressed gene products play a role in controlling embryo and endosperm development.

The biosynthesis of starch is the major determinant of yield in cereal grains. As ADPG is an activated sugar, the dependence of its production on respiration, changes which occur during development and the constraints which ATP production may place on carbon partitioning into different end products, are important.

Cereal grain development can be divided into two main stages, grain development and grain filling. The first stage, grain enlargement, involves early, rapid division of the zygote and triploid nucleus. Cell division is followed by the influx of water, which drives cell extension. This stage occurs at approximately 3–20 day post-anthesis. During the second stage (grain filling), cell division slows and then ceases and storage products are accumulated, beginning at around 10 days after anthesis until maturity, when the endosperm serves its function as a carbohydrate store.

Analysis suggests that 148 distinct proteins in addition to the highly abundant globulin and glutelin seed storage proteins. Identified proteins include those involved in RNA processing, translation, protein modification, cell signaling, and metabolism, as well as a number of hypothetical proteins.

Micro RNAs (mi RNAs) are upstream gene regulators of plant development and hormone homeostasis through their directed cleavage or translational repression of the target mRNAs, which may play crucial roles in rice grain filling and determining the final grain weight and yield.

Grain filling is divided into three stages viz. milk grain stage, dough grain stage and mature grain stage. In the milk grain stage, the grain has began to fill with a milky material. The grain starts to fill with a white, milky liquid, which can be squeezed out by pressing the grain between the fingers. The panicle looks green and starts to bend. Senescence at the base of the tillers is progressing. The flag leaves and the two lower leaves are green (8 days after fertilization). During the dough grain stage, the milky portion of the grain first turns into a soft dough and later into a hard dough. The grain in the panicle begins to change from green to yellow. Senescence of tillers and leaves is noticeable. The field starts to look yellowish. As the panicle turns yellow, the last two remaining leaves of each tiller began to dry at the tips (14 days after fertilization). In the mature grain stage, the individual grain is mature, fully developed, hard and has turned yellow. The upper leaves are now drying rapidly although. The leaves of some varieties remain green. A considerable amount of dead leaves accumulate at the base of the plant.

Blanking or spikelet sterility caused by poor anther dehiscence and low pollen production and hence low numbers of germinating pollen grains on the stigma is induced at this stage (Jagadish et al. 2007). Series of investigations have shown that spikelet sterility or blanking is induced by low temperature during the reproductive growth phase, especially during booting stage in areas with cool climate. Furthermore, Farrell et al. (2006) reported that low temperature during reproductive stage disrupts proper pollen development, leading to a shortage of sound pollen at the flowering stage.

It is noted, that spikelet fertility and specific spikelet fertility were mainly affected by the interaction effect between genotype and environment, with little maternal effect (Table 15). Interaction heritability in broad sense between genotype and environment for SF and SSF are 63.5% and 56.5% respectively, the highest compared with other effect values for heritability. Correlations between phenotype, genotype, additive and dominant effects of SF and SSF reached significant level with the correlation coefficients ranged from 0.717 to 1.000. Correlations between phenotype, genotype, additive and dominant effects of

Table 15 Spikelet weight and filling proportion of 12 'super' rice cultivars.

Cultivar	Grain weight (mg)			Filled grains (%)		
	Whole panicle	Superior	Inferior	Whole panicle	Superior	Inferior
Super rice						
Liangyoupeijiu	25.2	28.7	21.6	85.1	92.5	77.7
Xinliangyou	27.5	31.2	23.5	83.2	93.1	73.3
Ilyou 7954	27.4	30.9	23.6	84.9	94.3	75.5
Ilyon 084	27.1	30.4	23.4	85.6	95.2	76.0
Ilyonham 1	26.6	30.6	22.7	84.8	97.2	72.4
Ilyouming 86	27.2	30.5	23.8	87.3	97.5	77.6
Illyou98	27.0	30.3	23.5	83.9	93.8	74.0
Fengyou299	26.2	30.2	21.8	82.5	93.6	71.4
DYou 527	28.2	31.9	24.3	81.8	96.9	66.7
Wujing 15	28.9	31.8	24.9	86.4	96.5	76.3
Wujing1	26.7	29.4	23.8	85.5	94.9	75.9
Huaidaw9	27.2	31.5	23.9	87.5	97.9	77.1
Average	27.1	30.6	23.4	84.8	95.2	74.5
Conventional rice						
Shanyou 63ck	27.4	29.2	25.6	92.4	96.6	88.2
Yangdao 6ck	27.2	28.6	25.8	92.1	94.9	89.3
Yangfujing 8cK	26.7	28.1	25.3	95.8	98.2	93.4
Average	27.1	28.6	25.6	94.1	96.6	90.3

(after Yang and Zhang 2013)

SF and those of panicle exsertion (PE) also reached significant level with the correlation coefficients ranged from 0.161 to 0.975. The range of coefficient of variation varied with the traits, and that of PE were maximum compared with other three traits. Hence, PE was also a valuable trait in evaluation of cold tolerance. As for to PE and DH, their broad sense heritabilities were 59.6% and 81.4% respectively, the biggest compared with other effect values for heritability. DH was mainly controlled by genetic effects, with the least influence by environmental conditions. The low temperature stress should be one of the essential conditions to breed varieties with cold tolerance (Furong et al. 2009).

Results also showed a large spatial variability in terms of spikelet sterility within the field which related to water and air temperatures as well as their differences (Ortega et al. 2009).

Grain filling starts from apical part of a panicle and moves downwards in the main panicle and then in the branches gradually following the same process. In this way the green spikelets accumulate the translocated reserve assimilates from leaf sheath and culms.

To increase the yield further and to break the yield ceiling, breeding efforts have expanded the yield sink capacity mainly by increasing the number of spikelets per panicle. As a result, cultivars with large panicle have become available, such as the New Plant Type of the IRRI, and hybrid rice and 'super' rice or 'super' hybrid rice in China. These cultivars, however, do not frequently exhibit their high yield potential due to their poor grain-filling, as in a slow grain-filling rate and many unfilled grains (Ao et al. 2008).

Mechanism of grain filling is related to

i) Assimilate supply in relation to grain-filling.
ii) The enzymatic activity and molecular mechanisms involved in grain filling.
iii) Post anthesis moderate soil drying enhances grain filling.

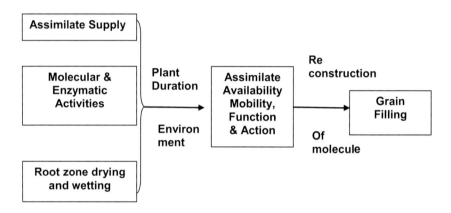

Under cold the sensitive varieties of rice have more inferior spikelets, often more than superior spikelets. As the number of superior spikelets are more, the variety is better adopted and called tolerant variety.

Rice grains at the apical primary branches of the panicle are known as 'superior', while those at the proximal branches are classified as 'inferior' (Wer et al. 2011). Cold tolerance evaluation of rice at the booting stage in different places and the correlationship between spikelet fertility and specific fertility (spikelet fertility of nine spikelets from the 3rd to the 5th spikelets of three primary branches at the top of each panicle) of single plant revealed a significant correlationship between spikelet fertility and specific spikelet fertility with the range from 0.738 to 0.901. However, under the cold stress environments, the correlation coefficients were higher than those without cold stress. On the other hand, the heredity analysis suggested that specific spikelet fertility could replace spikelet fertility as an identification indicator, while panicle exsertion could be used as a partial indicator of cold tolerance identification at the booting stage in rice (Dai et al. 2002).

The percentage of filled grains depends on the grain filling rate and grain filling duration of superior and inferior grains, which may be fast synchronous, slow synchronous or asynchronous. Grain weight of superior grains followed 'S' curve with faster filling rate, while that of inferior grains continuously increased during the filling stage linearly (Fig. 33).

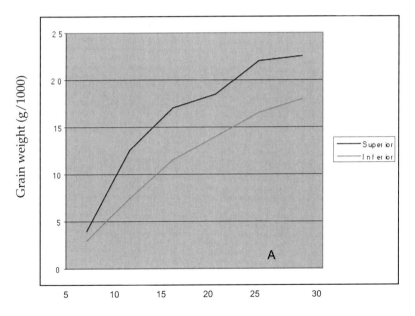

Fig. 33 contd....

Fig. 33 contd.

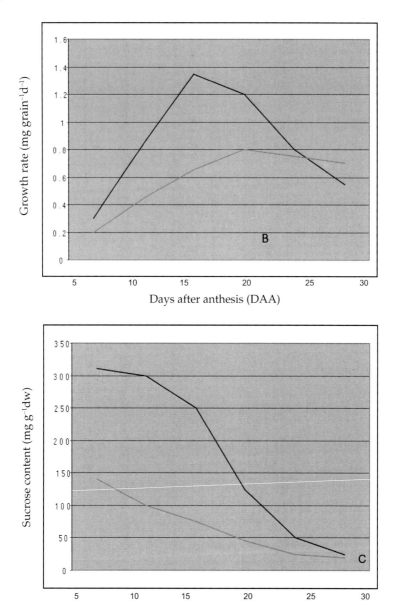

Fig. 33 contd....

Fig. 33 contd.

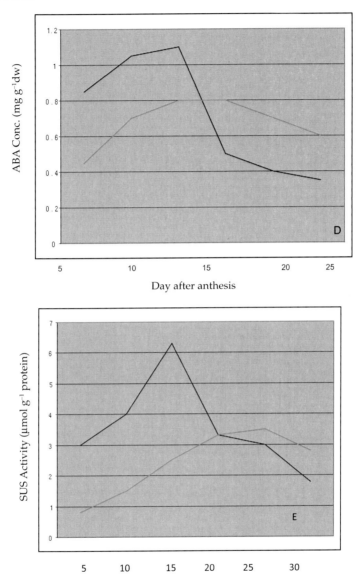

Fig. 33 Grain weight (A), Growth rate (B), Sucrose content (C), ABA Concentration (D) and SUS activity (E) changes during grain filling in superior and inferior grains.

Color image of this figure appears in the color plate section at the end of the book.

Rice plants grown under low temperature during reproduction phase and at flowering, often showed a similar trends of changes as that of inferior grains during grain filling period in rice.

Under cold stress at the reproductive phase, the grain sterility increases. Even the grain thus formed are either half filled or three fourth filled in irregular shape thus decreasing average grain weight. This might be due to the fact that cold or low temperature in post anthesis period severely affecting the translocation of reserve assimilates to the developing grain. Thus the accumulated assimilates like soluble carbohydrates and soluble nitrogenous compounds are being hindered to partition from source to sink (Annomymous 1989). Also, studies have shown that as the temperature decreases in hilly areas, translocation of soluble sugar and phosphorus to the developing grains gets affected adversely resulting in chaffy grains (Ngachan et al. 2010).

Increased yields have been achieved by (1) increased or extended photosynthesis per unit land area and (ii) increased partitioning of crop biomass to the harvested product. Increasing assimilate allocation to the reproductive primordial so as to establish a large potential sink should also increase total crop photosynthesis indirectly. Evidence in the major grain crops suggests that by anthesis the capacity of photosynthesis and that photosynthesis is not limiting during grain filling. To use this surplus capacity it is suggested that carbon and nitrogen partitioning to the reproductive meristem be increased so as to establish a high potential grain number and the potential for a large grain size. It is then expected that additional photosynthesis will follow, either by a longer daily duration of photosynthesis or by an extended leaf area duration (Richards 2000).

Several complex and interlinked processes are involved in cereal seed development. The starchy endosperm and aleurone layer are formed and storage proteins, lipids and polysaccharides are deposited in the endosperm. The mature seeds contain only about 10–15% water. The starchy endosperm cells do not survive this desiccation and undergo programmed cell death, whereas the aleurone and embryo stay alive but maintain a basal level of metabolic activity. Some proteins were present throughout development (e.g., Cytosolicmalate dehydrogenase) whereas others were associated with the early grain filling (ascorbate peroxidase) or desiccation stages. Most noticeably, the development process if characterized by an accumulation of low molecular weight compounds, α-amylase/trypsin inhibitors, serine protease inhibitors and enzymes involved in protection against oxidative stress (Finnie et al. 2002).

Low temperature may inhibit the photosynthetic rate through the inactivation of enzymes in chloroplasts as well as stomatal closure coupled by reduced water uptake. The photosynthesis and transpiration of the rice plant, as shown in Table 16 were considerably reduced by chilling at 5°C for 19 hrs, although no signs of chilling injury such as discolouration and wilting were observed.

Table 16 Influence of low temperature on photosynthesis and transpiration of rice plant (Tanaka and Yoshitomi 1973).

Before chilling				After chilling					
				3 hrs		4.5 hrs		12 hrs	
Variety	02%	P	T	P	T	P	T	P	T
Nankai 43	21	13.4	3.2	3.6	1.7	1.0	1.4	11.3	2.2
	03	17.9	3.0	0.9	1.5			17.8	2.4
IR 8	21	14.5	3.3	1.2	1.6	0.7	1.4	3.7	2.1
	03	18.3	3.0	0.2	1.4			7.3	2.2

P = Photosynthesis (mgco$_2$ dm^{-2}, hr^{-1})
T = Transpiration (g.dm^{-2}hr^{-1})

Before chilling 30°C, chilling at 5°C of flag leaf at milky dough stage and 50 Klux.

Mature storage tissues such as cereal endosperm consist primarily of starch and a minor pool of soluble sugar. Consequently, the major fate of translocated photosynthate entering the developing endosperm is to be metabolized by invertase or sucrose synthase into precursors for starch biosynthesis. The endosperm sucrose pool represents an alternative storage from the increasing photosynthate. It is unclear whether endosperm sucrose accumulation depends on transport of unmetabolized sucrose into endosperm cell or whether resynthesis of sucrose occurs from a pool of hexoses and nucleotide sugars common to the starch biosynthesis pathway (Smyth and Prescott 1989).

Resynthesis of sucrose might be catalyzed by sucrose synthase and sucrose phosphatase activities. The presence of sucrose phosphate synthases in many storage organs suggests that some sucrose may be resynthesized using the nonreversible pathway.

ADP glucose pyrophosphorylase (AGP) and sucrose synthase (SS) are important enzymes in starch biosynthesis pathway of rice endosperm (Fig. 34). It is hypothesized that difference in individual grain weight among rice cultivars or within a cultivar is directly related to the variation in activities of these enzymes and hormonal manipulation of enzyme activities can enhance grain yield. One potential mechanism for yield increases is that reduction of ethylene concentration at anthesis which may improve assimilate partitioning to grains and increases its weight. In an experiment, three indica rice cultivars differing in grain size and weight were grown in the field conditions during dry season. Dry matter growth, rate of division of endosperm cells, starch and sugar concentrations

Simplified pathway of starch synthesis in rice grains.

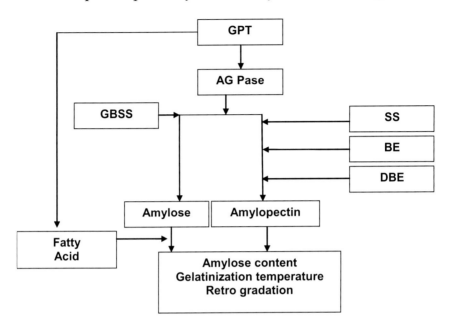

GPT = Glucose 6-phosphate translocator

AG Pase = ADP glucose pyrophosphorylase

GBSS = Granule bound starch synthesis

SS = Sucrose synthase

BE = Branching enzyme

DBE = Debranching enzyme

(after Kharabian-Masouleh et al. 2012)

as well as APG and SS activities of endosperm and ethylene evolution of spatially separated developing spikelets of panicle of the three cultivars were measured during the early part of grain filling period. Growth and cell division rates as well as activities of enzymes were higher in a big sized seed compared to small sized seed in sequel to the difference in either cultivars or positions on the panicle axes. Growth and enzyme

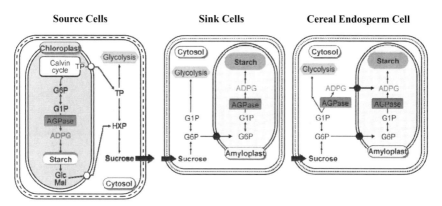

Fig. 34 Starch formation in grains.

activities correlated negatively with ethylene concentration. It is concluded that seed weight of inferior spikelets can be improved in rice panicle by increasing activities of starch synthesizing enzymes through manipulation of ethylene production, while genotypic difference in seed weight remains beyond manipulation (Mohapatra et al. 2009).

Activities of AGPG, pyrophosphorylase, soluble starch synthase and starch branching enzymes increased first and then fell down during grain development. Different varieties had different enzymes activities at the same grain filling stage and differed in the time when enzyme activity reached a peak value. Activities of ADPG pyrophosphorylase and soluble starch synthase respond dully to the temperature but starch branching enzyme was more sensitive to the temperature and its activity declined when temperature was too high or too low (Zheng-xun et al. 2005).

Starch synthesis in amyloplast in developing grain of rice may be given below. It suggests that sucrose transported is the starting material in association with sucrose synthase (SS) undergoes change to hexose phosphates and biphosphates.

Calvin cycle trioses are also converted into hexoses. Glucose phosphate subsequently polymerises into starch in association with ADPG Pase (Fig. 35).

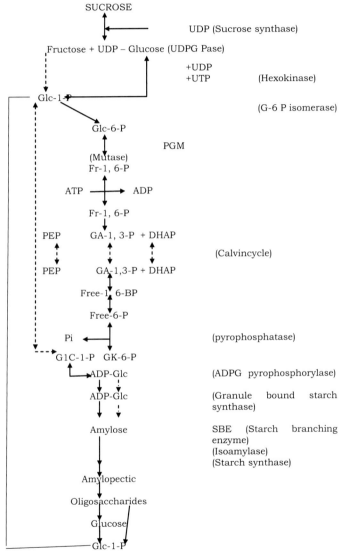

Fig. 35 Starch synthesis pathway in developing rice grain.

The relationships between the rate of starch accumulation and the activities of enzymes involving starch biosynthesis in developing grains of field grown rice cultivar Tainung 67. The results indicated that the activities of most of the grain showed their maximum between 7 to 14 days after anthesis, but the sugar-starch conversion rose and reached their maximum. The changes in the rate of starch accumulation correlated well with the changes, in the activities of sucrose synthase (SuS), invertase, hexokinase, soluble starch synthase (SSS), granule bound starch synthase, pyrophosphorylase, UDP glucose pyrophosphorylase, phosphoglucoisomerase, phosphoglucomutase, AGP glucose pyrophosphatase, starch debranching enzyme, during the grain filling period. The rapid grain-fill and shorter grain filling were associated with its higher activity of starch synthesizing enzymes of the short duration rice variety at the early phase of grain development. It is noted that starch accumulation is significantly related to invertase (0.52), hexokinase (0.54), phosphogluco mutase (0.56), AG Pase (0.42), soluble starch synthase (0.67) and starch branching enzyme (0.58) significantly (Table 17).

Leaf sheaths of higher position leaves (upper leaf sheaths) on rice (*Oryza Sativa* L.) stems function as temporary starch storage organs at the pre heading stage. Starch is quickly accumulated in upper leaf sheaths before heading, but the storage Starch is degraded at post heading stage to provide the carbon source for developing grains. Abscisic acid (ABA) is a key plant hormone to control plant development and stress responses. It was found that ABA content in upper leaf sheaths was significantly increased at the stage after panicle exsertion and that the pattern of ABA increase was negatively correlated with changes in starch content. Exogenous ABA reduced starch content in leaf sheaths while the activities of starch degradation

Table 17 The correlation coefficients (r) between dry matter accumulation as well as starch accumulation with sucrose to starch conversion enzymes in rice CV. Tainung 67.

Parameter rate	Drymatter Accumulation	Starch Accumulation
Starch accumulation rate	0.99**	--
SUS	0.41*	0.19
Invertase	0.42*	0.52**
Hexokinase	0.76**	0.54**
Phosphoglucoisomerase	0.21	0.18
Phosphoglucomutase	0.65**	0.56**
AG Pase	0.64**	0.42**
UD Pase	0.37	0.27
SSS	0.73**	0.67**
GBSS	0.02	0.07
SBE	0.69**	0.58**
SDBE	0.23	0.12
Soluble protein	0.62**	0.52**

(after Tseng et al. 2003)

enzymes (viz. a-amylase, b-amylase) increased in ABA treated leaf sheaths and fructose transporter gene expression was up regulated. However, ABA repressed the activities of some starch biosynthesis enzymes (i.e., ADP-glucose pyrophosphorylase, granule bound starch synthase) in leaf sheaths. These results suggest that ABA plays an important role in promoting starch degradation and sucrose remobilization in upper leaf sheaths at the post heading stage.

The degradation of storage proteins happened mainly at the late stage of germination phase II (48hr imbibition). In addition to alpha-amylase, the up-regulated proteins were mainly those involved in glycolysis such as UDP-glucose dehydrogenase, fructokinase, phosphoglucomutase, and pyrurate decarboxylase.

Prior to the synthesis of amylose and amylopectin, sucrose loaded into endosperm cells by sucrose transporters and converted to Glc-6-P via several reaction steps catalyzed by enzymes such as Suc synthase and then into ADP-G1C, the substrate for α-1, 4-polyglucan synthesis, by ADP-G1C pyrophosphorylase. Import of ADP-G1C and G1C-6-P into amyloplasts is conducted by distinct transporters. For each enzyme and transporter involved, several isoforms are known to be differentially regulated. Although amylose synthesis is exclusively governed by granule bound starch synthase, amylopectin is synthesized via concerted reactions catalyzed by soluble starch synthase, branching enzyme and starch debranching enzyme (Nakamura 2002). In particular, the relative balance of α-1, 6 branch formation and the subsequent α-1, 4-chain elongation which are catalyzed by distinct branching enzyme and soluble starch synthase isoforms, respectively, are important for determining the amylopectin fine structure (Nakamura 2002) (Fig. 36).

A comprehensive analysis of the transcript levels of genes which encode starch synthesis enzymes is fundamental for the assessment of the function of each enzymes and the regulatory mechanism for starch biosynthesis in source and sink organs. Using quantitative real-time RT-PCR, an examination was made of the expression profiles of 27 rice genes encoding six classes of enzymes, i.e., ADP glucose pyrophosphorylase (AG Pase), starch synthase, starch branching enzyme in developing seeds and leaves. The modes of gene expression were tissue and developmental stage specific, four patterns of expression in the seed were identified: group I genes, which are expressed very

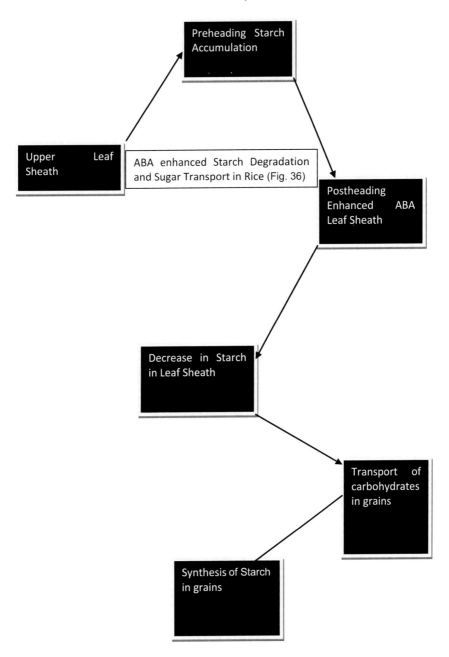

early in grain formation and are presumed to be involved in the construction of fundamental cell machineries, *de novo* synthesis of glucose primers, and initiation of starch granules, group 2 genes, which are expressed highly throughout endosperm development, group 3 genes, which have transcripts that are low at the onset but rise steeply at the start of starch synthesis in the endosperm and are throughout to play essential roles in endosperm starch synthesis, and group 4 genes, which are expressed scantly, mainly at the onset of grain development and might to be involved in the synthesis of starch in the pericarp. It is also revealed that the defect in the cytosolic AG Pase small subunit 2b (AGP2b) transcription from the AGPS2 gene in endosperm sharply enhanced the expressions of endospermus and leaf plastidial AGPS1, the endosperm cytosolic AG Pase large subunit 2 (AGPL2), and the leaf plastidial AGPL1 (Ohdan et al. 2006).

Temperature during grain filling, planting year, kind of seedling, heading date and the location of grains on panicle were analyzed in Hokkaido a cold district. A significant negative correlation was found between temperature during grain filling and the amylose content (Table 18).

Temperature influenced the amylose content greatly when the accumulated temperature after heading was under 800°C. A significant negative correlation was found between seedling age and amylose content. The s.d. of amylose contents of grains from panicles of the same heading date was 0.38–1.34. The amylose content differed with the location of grains. It was higher in the grains in a higher position and are higher than that in a lower position. The amylose content of grains on the primary rachis (part of panicle) branch was higher than that on secondary rachis branches. This is probably because the starch content of the grains on the secondary rachis branches

Table 18 Range of physiochemical properties of 233 Australian rice grain genotypes.

Trait	Range (Unit)
Peak	2168–3669
Tough1	1312–2372
Breakdown Viscosity	667–1913
Final Viscosity	2560–4386
Set back	−658– +1203
Peak time	5.7–6.3
Pasting temperature	65.65–78.40
Martin test (N)	0.405–3.612
Amylose content (%)	14.10–28.85
Predicted N (N)	0.31–1.82
Gelatinization Temperature	62.00–82.98
(°C)	0.709–44.55
Chalkiness (%)	

was lower than that on the primary rachis branch. On the other hand, the rate of milk white grains and white belly grains in higher position was lower than that of those in a lower position. Moreover, the rate of milk white grains and white belly grains on the primary rachis branch was lower than that on the secondary rachis branches. The rate of milk white grains and white belly grains was negatively correlated with amylose content. It is concluded that reducing the number of grains on the secondary rachis branches is important to improve quality (Igarashi 2008).

Most amino acids including 4-amino butyrate (GABA) a tripeptide, glutathione in both oxidised and reduced forms, and amines such as putrescine increased in response to low temperature, including proline, etc.

Tolerance to abiotic stress was observed for over production of proline. Proline seems to have diverse roles under osmotic stress conditions, such as stabilization of proteins, membranes and subcellular structures, and protecting cellular functions by scavenging reactive oxygen species (Kishor et al. 2005).

Endosperm protein of rice during seed development confirmed that storage protein begins to accumulate about 5 days after flowering. Two polypeptide groups, 22 to 23 and 37 to 39 KDa, the components of glutelin, the major storage protein in rice seed, appeared 5 days after flowering. A 26 KDa polypeptide, the globulin component, also appeared 5 days after flowering. Smaller peptides (10 to 16 KDa) including prolamin components, appeared about 10 days after flowering. In contrast, the levels of the 76 and 57 KDa polypeptides were fairly constant throughout seed development, the accumulation of glutelin and globulin, was found faster than accumulation of prolamin (Yamagata et al. 1982).

Protein	Days after flowering
Glutelin	5
Globulin	5
Prolamin	10

It is also noted that the levels of fertilizer application in the cultivation influenced the protein and amylose contents in rice varieties appreciably. In an experiment with six rice cultivars it is observed that high fertilizer enhanced significantly the protein content, in some varieties it had been doubled, on the other hand amylose content decreased significantly. Thus, there is a negative correlationship between grain protein and amylose content. IR71137–243 has high protein content and IR72 has high amylose content (Table 19).

Table 19 Protein and apparent amylose contents of rice cultivars with low and high fertilizers (LF & HF).

(IRRI)

Cultivar	Protein LF	Protein HF	Amylose LF	Amylose HF
Basmati 370	7.2 ± 0.0	12.2 ± 0.3	23.5 ± 0.2	21.8 ± 0.2
IR60	6.5 ± 0.2	12.9 ± 0.5	29.0 ± 0.4	26.2 ± 0.4
IR71137–243	7.4 ± 0.1	12.3 ± 0.5	26.6 ± 0.3	24.1 ± 0.3
IR72	5.4 ± 0.3	11.0 ± 0.0	29.2 ± 0.4	26.9 ± 0.3
IR8	5.9 ± 0.3	9.7 ± 0.5	28.7 ± 0.2	27.1 ± 0.2
IR841–85	6.2 ± 0.1	11.1 ± 0.0	20.3 ± 0.6	19.7 ± 0.5
Mean	6.4 ± 0.7	11.6 ± 1.0	26.2 ± 3.4	24.3 ± 2.8

(after Champagne et al.)

Fertilizer level (Nitrogen) affects both protein and apparent amylose contents of rice grains. Interestingly the protein content of the low nitrogen treatment was within normal ranges of 6–7.5% (Shih 2004) for almost all cultivars. For each cultivar protein contents were significantly (P <0.0001) higher at the higher fertilizer level, whereas apparent amylose contents were significantly (P <0.0001) lower. This suggests that protein synthesis is stronger than starch synthesis, which is consistent with data indicating that mechanisms within the endosperm exist to partition the supplied substrates towards the maintenance of protein synthesis at the expense of starch production (Emes et al. 2003).

Two factors affecting protein accumulation during the first 2 weeks of grain development—the level of free amino acids and the capacity of the intact grain to incorporate amino acids—are correlated with the varietal differences in protein content of the mature rice grain. High protein grains have higher levels of all the nitrogenous fractions including RNA.

The effect of the level of free amino acids on protein accumulation is quite obvious. Amino acids are the precursors of storage proteins in the rice grain. Hence, a higher level of these amino acids will contribute to faster and greater accumulation of protein in the so called protein bodies (PB), even if the level of enzymes involved in protein synthesis are the same in all grains.

The greater capacity for amino acid incorporation in the developing grain implies a higher concentration of enzymes involved in protein synthesis.

The low activities of the RNAase and protease and their lack of relation to the accumulation of protein in the ripening grain suggests that the equilibrium between synthesis and degradation is shifted toward the former.

Rice storage proteins (prolamin) that accumulate in PB-1 appear to be synthesized by membrane—bound polysomes attached to PB-1 or RER and to pass through the membrane into lumen where they aggregate and are deposited. The proteins (glutelin and globulin) that accumulate in PB-II, however, seem to be synthesized by membrane—bound polysomes as a large precursor and to become sequestered into the cisternal space of RER, from where they are transferred to vacuolar precursor of PB-II.

In japonica rice varieties, the activities of glutamine synthetase gradually increased and then declined as a single peak curve in the course of grain filling. The 15th day after heading was a turning point, before which the enzymatic activities in the inferior rice varieties with high protein content were higher than those in the superior rice varieties with low protein content and after which it is reversed. It was correlated with the protein content of rice grain and set back positively at the early stage and negatively at the middle and late stages. The correlation degree varied with the course of grain filling. From 15 to 20 days after heading was a critical stage (Jin et al. 2007) (Tables 20, 21 & 22).

Table 20 Multiple comparison of glutamine synthetase in rice grain filling.

Glutamine synthetase activity (OD/grain. h)					
Variety	10DAH	15DAH	20DAH	25DAH	30DAH	35DAH
Shuiludao 1	0.0271	0.0301	0.0243	0.0124	0.0143	0.0115
Toukei 180	0.0156	0.0211	0.0255	0.025	0.0187	0.0151
Fujihikan	0.0184	0.0224	0.0252	0.0202	0.0160	0.0126
Dongnong 415	0.0202	0.0241	0.0199	0.0160	0.0128	0.0102

Table 21 Multiple comparison of nitrogen contents in rice grain filling.

Nitrogen content in rice grain (mg/grain)					
Variety	10DAH	15DAH	20DAH	25DAH	30DAH	35DAH
Shuiludao 1	24.87	33.69	37.67	41.10	42.36	42.22
Toukei 180	17.03	25.90	33.45	36.49	37.03	36.49
Fujihikan	20.57	29.16	35.06	37.57	38.28	38.10
Dongnong 415	22.19	32.20	35.86	38.42	39.71	39.71

Table 22 Correlation between glutamine synthetase activity and protein content taste meter below, RVA Properties at different grain filling stages.

Days after heading	Protein content	Tastemeter value	Peak viscosity	Breakcolour	Setback
10	0.4061	−0.9201	−0.6365	−0.7511	0.7614
15	0.3885	−0.9219	−0.5656	−0.6817	0.6894
20	−0.9610*	0.5047	0.7375	0.5899	−0.5237
25	−0.9204	0.7536	0.9598**	0.8943	−0.8546
30	−0.8616	0.7864	0.9870*	0.9493*	−0.9201
35	−0.8316	0.7512	0.9954**	0.9600**	−0.9347

The major storage proteins of most cereal grains are glutelin and prolamin. In rice grains, however, glutelin is the major protein of the starchy endosperm, constituting at least 80% of the total protein, prolamin accounting for less than 5%. The increase in glutelin in the developing rice grain coincide, with the appearance of Protein body (PB) in the starchy endosperm 7 to 8 DAF and glutelin is found exclusively in PB. In a study it had been reported that there are two types of PB in the starchy endosperm of rice grains. One PBI is spherical with a concentric ring structure, whereas the other PBII is stained homogeneously by osmium tetraoxide and does not have this structure. PBI contains prolamin and PBII is rich in glutelin and globulin.

Little, however, is known about the molecular mechanisms which control the synthesis of the storage proteins of rice seeds. Studies showed that storage proteins begin to accumulate about 5 days after flowering. The components of glutelin, the major storage protein in rice, appeared 5 days after flowering. Smaller peptides, including prolamin appeared about 10 days after flowering. PBII type protein grows faster than PBI type protein. Rice grains from 3 varieties and 3 pairs of lines with different protein content were collected at 4 days intervals from 4 to 32 days after flowering. Analysis for protein, free amino acid nitrogen, ribonucleic acid, protease and ribonuclease activities, in addition, the capacity of the intact grain to incorporate amino acids showed maximal level of free amino acid nitrogen and the capacity of the developing grain to incorporate amino acids were consistently found to be higher with high protein content grains (Fig. 37). The ribonucleic acid content of the grain also tended to be higher in high protein grains (Cruz et al. 1970). The rice with high percentage of protein tended to translocate more leaf N to the developing grains than the rice with average grain protein content. The leaf blades of the former had also

Grain Development

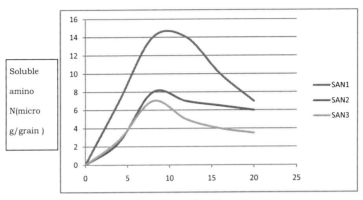

Soluble amino N(micro g/grain)

Days After Flowering

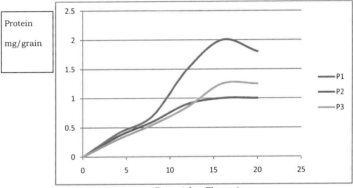

Protein mg/grain

Days After Flowering

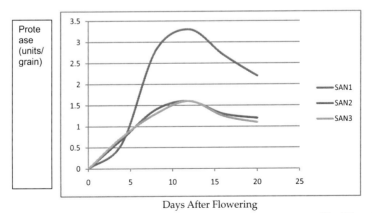

Protease (units/grain)

Days After Flowering

Fig. 37 contd....

Fig. 37 contd.

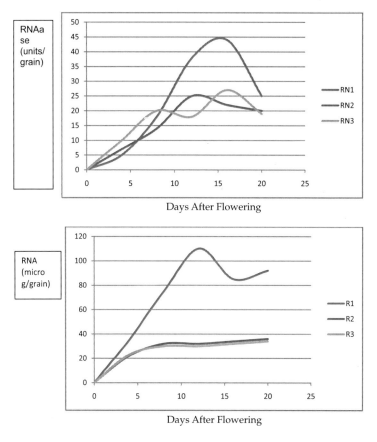

Fig. 37 Soluble amino N, Protein, Protease, RNase and RNA changes during grain filling in high protein and low protein rices.

Color image of this figure appears in the color plate section at the end of the book.

lower rates of leucine incorporation during grain development but higher protease activity than leaves of the rice with average protein content.

Nitrogen metabolism characteristics, such as glutamine synthetase activity, soluble protein content and free amino acids were known to be involved in the changes of protein concentration (Zhao and Yu 2005). Integrating the previous reports in wheat and other crops, Xie et al. (2003) concluded that enhanced ABA contents generally increase protein concentration, the relationship between the IAA contents and protein accumulation in grains was negative at the grain enlargement stage, but positive at the grain filling stage and the effects of enhanced ABA and IAA contents on protein accumulation were mediated mainly via enhancing mRNA production and protein gene expression. Thus, grain filling process and protein concentration were influenced by changes of endogenous hormone levels. Application of exogenous GA_3 in grain filling stage or even in earlier stage may be beneficial for grain filling in rice cultivation under low temperature, as low temperature enhanced ABA concentration, increased protein content in grain and tends to behave like inferior grains.

Rice grain filling is a process of conversion of sucrose into starch catalyzed by a series of enzymes. Sucrose synthase (SUS) is considered as a key enzyme regulating this process. Field grown rice plants and detached cultured panicles under varying treatments including spikelet thinning, leaf cutting and application of different concentrations of exogenous sucrose and ABA during grain filling indicates that a higher SUS activity was found in superior grains than in inferior grains in the earlier stage of grain filling, which was significantly and closely related to a higher grain filling rate and starch accumulation. An increase in sucrose concentration in grains as a result of different treatments increased both SUS activity and SUS protein expression in grains. An increase in ABA concentration gave similar results. Furthermore, effects of interactions between sucrose and ABA on the activity and expression of SUS protein in grains suggests that sucrose and

ABA mediated rice grains filling is largely due to increase in SUS activity and SUS protein expression (Tang et al. 2009).

Cold acclimation of plants is a highly active process resulting from the expression of a number of physiological and metabolic adaptations to low temperature (Levitt 1980). Major metabolic changes in carbohydrates, protein, nucleic acids, amino acids, and growth regulators have been documented during the acquisition of cold tolerance. Among these, water soluble carbohydrates like the fructose polymers and fructans were shown to accumulate during cold acclimatization of grass species (Livingston 1991). Fructans are claimed to enhance the cold tolerance in plants. The ability of plants to synthesize fructans correlates their survival in colder climates. Relationship between cold tolerance and fructan accumulation has been noted in cereals (Suzuki and Nass 1988, Pontis 1989). Evidence suggests that soluble sugars, such as sucrose and oligosaccharides of the raffinose family, in combination with heat stable proteins could play a determinant role in cold stress tolerance by protecting proteins and membranes against stress (Gusta et al. 1996).

Cold stress induced major changes in amino acid levels in over wintering crowns of the three ecotypes and the highest contributions to total amino acid accumulation after acclimatization at low temperatures came from proline, glutamine and glutaric acid. Heat stable proteins have been isolated from cold-acclimated plants and a correlation between a heat-stable protein accumulation and cold induced tolerance suggests that protein induced may act in combination with soluble carbohydrates. During cold acclimation, homologs of LEA proteins also accumulate in many plant species. During cold water is drawn out of the cells, resulting in cellular dehydration (Guy 1990). Thus, LEA protein homologs may play a role in conferring tolerance in plant cells under low temperature. Cold acclimatization-induced accumulation of cold regulated (COR) proteins, having structural similarity with LEA proteins have a highly hydrophilic feature and remain soluble and stable. LEA proteins have simple and a few amino acids especially hydrophilic proteins and share the unique property of heat stability. Hence, COR and LEA proteins induced by cold is important for the better performance of the crop.

Studies indicated that protein accumulation is related to grain filling process and protein concentration, which were regulated by environmental factors, such as water conditions, nitrogen application, temperature stress, etc. Plant growth regulators determine the level of proteins, which respond differently to the environmental stimuli (Bano et al. 1993, Jackson et al. 1988). Numerous studies proved that changes in endogenous hormone contents affected grain filling process and protein concentration under different cultivation practices (Majid et al. 2011). Grain filling process was closely associated with the changes in hormone levels (Yang et al. 2003). The increase in ABA contents at the end of grain filling and its rapid fall during maturation have leads to an assumption that ABA plays an important role in dry matter accumulation, GA_3 have been proved to be an important phytohormone that regulates the duration of grain filling (Zhang et al. 2009).

Plant hormones are considered as key regulators to seed development. Morris et al. (1993) reported that zeatin in developing rice grains showed large transient increases following pollination, which coincided with the period of seed setting and maximum endosperm cell division. There are many reports that auxins, gibberellins, and abscisic acid are also involved in regulating grain development (Hansen and

Grossmann 2000) stated that GA like material was highest in liquid endosperm. Kato et al. (1993) reported that ABA content in large-size grains was higher than that in small size grains during rice grain filling (Fig. 38).

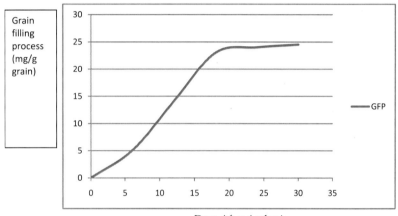

Days After Anthesis

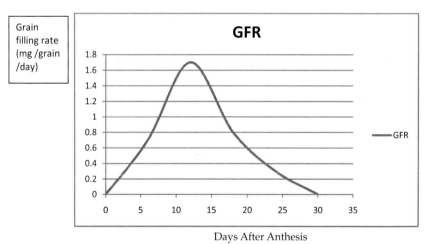

Days After Anthesis

Fig. 38 contd....

Fig. 38 contd.

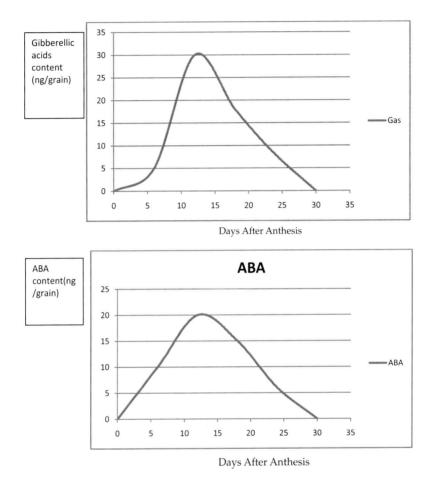

Fig. 38 Grain filling, rate of grain filling, GA and ABA changes in rice grain after anthesis.

Three grain filling patterns based on the filling rate of superior and inferior spikelets were observed, i.e., fast synchronous, all spikelets started filling early and fast at the early filling stages; slow synchronous; all spikelets filled slowly at the early filling stage and reached the maximum filling rate late; and asynchronous; superior spikelets started filling and reached the maximum filling rate much earlier than the inferior ones. The order of grain filling percentage in the three types of grain filling patterns was: fast synchronous > asynchronous > slow synchronous. Changes in zeatin and zeatin riboside contents in the superior and inferior spikelets were associated with the grain filling patterns. Grain filling percentage was significantly correlated with zeatin + zeatin riboside contents in the grains and roots at the early and middle grain filling stages. IAA and GA contents in the grains and roots were not significantly correlated with grain filling percentage. It is suggested that cytokinins in the grains and roots during the early phase of grain development play an important role in regulating grain filling pattern and consequently influence grain filling percentage.

Brassinosteriods (BRs) are phytohormones mediating multiple biological processes, such as development and stress response. They have been used in crops to produce high yield. In rice, the ideal plant architecture for high yield includes effective tillers, as well as height and leaf angle, which is modulated by BRs.

It had also been shown that endogenous hormones are essential regulators for grain filling in cereals and regulate grain weight via influencing grain filling process and activities of key enzymes (Yang et al. 2004). It has been noted that a post anthesis wet and moderate drying holds great promise to improve grain filling of inferior spikelet through elevating cytokinin in levels in the rice plant (Zhang et al. 2010).

Major biological processes such as cereal grain filling are believed to require a close coordination of gene expression among many important pathways. However, direct experimental evidence for this hypothesis has been lacking due to absence

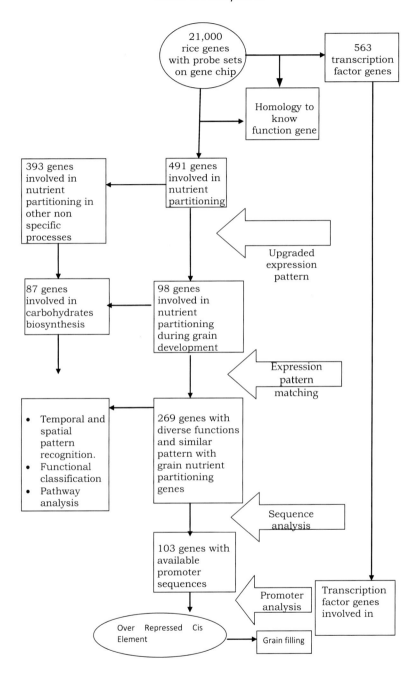

of genomic sequence information and functional genomic technologies. Through cluster analysis 98 genes are identified preferentially expressed during grain development. Schematic diagram of the data mining process when grain filling genes were selected based on homology to known function genes and their expression in 33 samples was determined during rice development.

6

YIELD AND YIELD COMPONENTS

Sustainable increase in rice production in the near future requires substantial improvement in productivity and efficiency. Rice yield and production have been considerably increased during the last 30 years. In a number of countries, yields of rice in favourable ecologies have reached the research yield potential of the present generation of high yield ceiling, where yield gaps are nearly closed. These increases not only enhance rice productivity but also the efficiency in production systems, resulting in high economic output as well as high income of farmers.

On the other hand, in many countries, the gap between yields at research stations and in farmers fields are still substantially large due to a combination of lack of initiatives, resources and goodwill to narrow them. In these countries the integrated crop management approaches including available location-specific technologies coupled with active institutional support from governments particularly for input and village credit supplies as well as stronger research and extension linkages, can expedite the bridging of yield gaps, thus improving the productivity and efficiency of rice production (Dunayri et al. 1998) (Table 23).

Table 23 Comparative national average yields, irrigated rice yields and experimental station rice yields, Asian countries, 1991.

Countries	National average rice yield (t/ha)	Irrigated rice yield (t/ha)	Average Potential rice yield (t/ha)
Bangladesh	2.6	4.6	5.4
China	5.7	5.9	7.6
India	2.6	3.6	5.9
Indonesia	4.4	5.3	6.4
Nepal	2.5	4.2	5.0
Myanmar	2.7	4.2	5.1
Philippines	2.8	3.4	6.3
Thailand	2.0	4.0	5.3
Vietnam	3.1	4.3	6.1

Rice production is affected severely due to cold from place to place and year to year, ranging from 10 to 50%. The yield reduction is mainly due to reduction of one of the main yield components namely the spikelet fertility which minimizes the number of filled grains in the spike. It is also noted that the other yield components such as panicle number per m² and the panicle length were affected due to panicle exsertion problems. Grain filling is also affected, as there are partially filled grains reducing the average 1000 grain weight. It is well known that the relationship of yield and yield components for rice may be given as Yield (g per m²) = panicles per m² X number of spikelets per panicle x grain fertility % x 1000 grain weight(g) X 10^{-5}. The yield components are often positively and negatively correlated with the yield depending on the situation. It is well known that japonica varieties are more tolerant to cold than those of indica varieties. Japan has more rice productivity than the average world rice productivity. The high yield is due to the genotype characteristics related to nitrogen responsiveness and ecological prevalent conditions. The average rice productivity in Japan is about 7 tons per hectare and that of other part of world is 4 tons per hectare (Fig. 39).

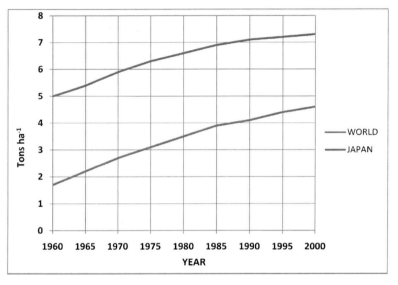

Fig. 39 Yearwise rice productivity in Japan and World.

Color image of this figure appears in the color plate section at the end of the book.

To understand the climatic influences on yield and yield components of rice under seasonal variation at different locations of tropics, IRRI carried out a study and analyzed the data. Yield varied from 3.49 to 11.60 tons per hec. Yield components also varied greatly at different locations suggesting that rice yield potential is high and can be achieved under proper management system and environmental conditions (Table 24).

Table 24 Seasonal variation in yield and yield components at four locations in the tropics.

Location	Planting Time	Number of crops	Variety	Yield range (t/ha)	cv
Los Banos	Jan-Dec	19	IR747-B2-6	4.21-7.13	15.4
Bankhen	Jan-Dec	11	RD1	3.49-5.77	13.2
	Jan-Dec	11	IR8	3.70-6.13	13.8
Chinat	Dec-July	8	RD3	5.11-7.28	12.5
	Dec-July	8	RD4	4.67-7.09	13.6
	Dec-July	8	C4-63	5.33-6.63	9.0
	Dec-July	8	IR661-140	5.61-7.64	9.4
Hyderabad	July & Jan	2	IET1991	7.6 & 11.6	-
	July & Jan	2	IET1039	6.1 & 11.1	-
Spikelet number (10^3/m^2)					
Los Banos	Jan-Dec	19	IR747-B2-6	29.0-47.0	15.1
Bangkhen	Jan-Dec	11	RD1	21.3-28.4	7.9
	Jan-Dec	11	IR8	22.8-29.1	6.9
Chinat	Dec-July	8	RD3	18.1-28.9	12.8
	Dec-July	8	RD4	23.0-32.6	12.3
	Dec-July	8	C4-63	22.0-33.0	12.7
	Dec-July	8	IR661-140	23.3-35.8	13.8
Hyderabad	July & Jan	2	IET1991	57.2 & 66.1	-
	July & Jan	2	IET1039	57.1 & 81.7	-
Grain Weight (g/1000)					
Los Banos	Jan-Dec	19	IR747-B2-6	17.2-19.1	2.9
Bangkhen	Jan-Dec	11	RD1	27.2-28.6	5.7
	Jan-Dec	11	IR8	27.2-30.1	3.1
Chinat	Dec-July	8	RD3	29.5-32.6	3.8
	Dec-July	8	RD4	29.4-32.2	2.9
	Dec-July	8	C4-63	24.3-27.3	3.7
	Dec-July	8	IR661-140	24.7-28.1	3.8
Hyderabad	July & Jan	2	IET1991	20.3 & 20.3	-
	July & Jan	2	IET1039	18.3 & 18.3	-

Table 24 contd....

Table 24 contd.

Location	Planting Time	Number of crops	Variety	Yield range (t/ha)	cv
			Filled Grain (%)		
Los Banos	Jan-Dec	19	IR747-B2-6	83.0-95.2	4.0
Bangkhen	Jan-Dec	11	RD1	66.3-83.5	7.5
	Jan-Dec	11	IR8	58.4-76.2	9.3
Chinat	Dec-July	8	RD3	91.7-96.1	1.7
	Dec-July	8	RD4	80.0-96.5	8.2
	Dec-July	8	C4-63	87.0-92.7	2.3
	Dec-July	8	IR661-140	90.7-94.5	1.4
Hyderabad	July & Jan	2	IET1991	61.0 & 94.3	-
	July & Jan	2	IET1039	59.5 & 82.1	-

Source: IRRI

According to AICRIP, India, grain size showed negatively linear relationship with number in early duration group of varieties. The decrease in grain number with increased size was mainly due to the decreased number of grains per panicle. The grain number recorded was of 25×10^3 which was lower in comparison to the grain number recorded in medium duration and late duration group varieties (Fig. 40). Grain size was found to be not related to grain yield. Apparently increased grain number at any given size would result in higher yields. Suitable blending of grain size and number would lead to further increases in yield potential. At a given grain size grain yields differed suggesting greater scope for manipulations. Based on the grain filling pattern in terms of growth rate and peak filling periods, the varieties were grouped with peaks around 10, 15, 20 and 25 days after flowering. The grain size was found not to be related to grain filling periods and growth rate of grains but on the climatic conditions (Rao et al. 1984).

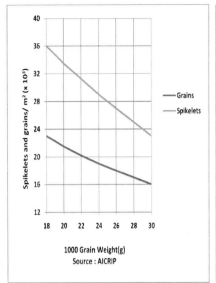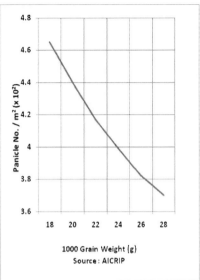

Fig. 40 Relationships of spikelets, grains and panicles per m² with 1000 grain wight.

Color image of this figure appears in the color plate section at the end of the book.

Apart from abiotic stress, yield and yield components are also influenced by the level of fertilizer, i.e., by management practices. According to Fageria et al. (2011) the yield and yield components increased along with the application of nitrogen level upto a certain point and declined gradually thereafter. Therefore, on the basis of yield and yield component data, the optimal level of nitrogenous fertilizer application can be ascertained. Higher level of nitrogen increases the grain sterility percentage appreciably (Tables 25 & 26).

In the same situation the relationship between filled grain percentage showed a negative relationship with the concentration of nitrogen in the leaf as percentage. Therefore,

Table 25 Yield and yield components of upland rice as influenced by nitrogen application.

Nitrogen Rate (mgKg⁻¹)	H1	Grain Sterility (%)	1000-grain Weight	Panicle (g) per plant	Grain yield (g plant⁻¹)
0	0.48	13.22	28.58	2.52	2.51
50	0.51	14.55	28.93	3.40	6.82
100	0.53	16.58	29.49	3.69	10.21
150	0.54	18.84	29.70	4.10	13.50
300	0.53	13.27	28.31	4.20	15.00
400	0.51	20.14	27.38	4.00	14.00
Mean	0.52	16.10	28.73	3.65	10.34

** CV.BRS sertaneja

Table 26 Correlation coefficients between grain yield and yield components irrespective of nitrogen level.

**	Parameter	Correlation coefficient
**	Panicle number	0.819**
	Grain sterility (%)	0.143
	Harvest Index	0.260*
	1000 grain weight (g)	0.085

** and * significant at 1% and 5% levels respectively

too high nitrogen application or concentration of nitrogen in the leaf is associated with the reduction on filled grains (Fig. 41).

Similarly, application of phosphorus influenced the yield and yield components in rice. A field experiment, at Chungbuk, showed that there are appreciable changes due to application of phosphorus; all yield and yield components are enhanced by the application of phosphorus, even the grain filling ratio (Table 27).

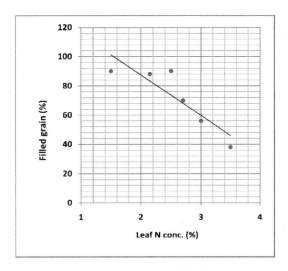

Fig. 41 Relationship between percentage filled grain and leaf nitrogen concentration.

Table 27 Effect of phosphate application on rice under low temperature.

Phosphate	Grain	Filled grain	Milled rice
(Kgha⁻¹)	(No.in 10³m⁻²)	ratio (%)	yield (tha⁻¹)
**			
100	27.5	87.0	3.89
150	28.6	89.0	4.16
200	31.1	90.4	4.49

Source: Chungbuk RDA (1980)

Temperature regimes greatly influence not only the growth duration but the growth pattern of the rice plants. Rice varieties from Japan which has low temperatures during the cropping season when grown in Indonesia which has moderate to high temperatures during the cropping season, for example, generally mature early whereas Indonesian varieties when grown in Japan generally extend growth duration. In temperate countries, generally, low temperature regimes limit rice cropping to only one season. On the other hand, it is well known that respiration is low at low temperatures. The low respiration due to low night temperatures during the grain development phases of rice plants planted outside of the TCTC zone may favour grain development and filling. This high yield (Table 28) provider excel the critical temperatures for different phases of rice plants.

Table 28 Critical temperatures at different growth stages of rice plants (adopted from Yoshida 1978).

Growth State	Critical temperature (°C)		Optimum temperature (°C)
	Low	High	
Germination	16–19	45	18–40
Seedling emergence	12	35	25–30
Rooting	16	35	25–28
Tillering	9–16	33	25–31
Panicle initiation	15	-	-
Panicle differentiation	15–20	30	-
Anthesis	22	35–36	30–33
Ripening	12–18	>30	20–29

The optimum planting time of transplanted aman in Bangladesh is August. But delay in transplanting in late planting exposes the reproductive phases as well as phenological events of crop in an unfavourable temperature regime thereby causing high spikelet sterility and poor growth of the plant (Nahar et al. 2009).

Effects of low temperature stress in transplanted aman rice varieties mediated by different transplanting date are noted on the yield and its components (Tables 29 & 30).

Table 29 Changes in yield components with the day in transplanting.

Date of transplanting	No. of Panicle per hill		No. of Filled grains per panicle		Spikelet sterility (%)		1000 grain wt.(g)	
	BD 46	BD 31	BD 46	BD 31	BD 46	BD 31	BD 46	BD 31
Sep 01	16.5	15.5	125	120	13.5	11.5	26.0	27.0
Sep 10	17.0	13.0	115	105	14.5	15.5	26.0	26.5
Sep 20	14.0	12.0	110	90	15.4	17.4	23.5	23.0
Sep 30	12.5	8.0	105	81	17.0	24.0	23.0	20.0

CV. BD46 and BD31

Table 30 Yield and harvest index of rice varieties as affected by transplanting dates.

Date of transplanting	Grain yield (tha⁻¹)		Straw yield (tha⁻¹)		Harvest index (%)	
	BD 46	BD 31	BD 46	BD 31	BD 46	BD 31
Sep 01	4.50	4.90	7.60	9.80	37.19	33.33
Sep 10	4.30	4.60	8.25	9.76	34.26	32.03
Sep 20	4.10	3.90	7.54	7.65	35.22	33.77
Sep 30	3.80	3.10	6.70	6.40	36.19	23.63

The Korean peninsula is located in the Far East, between latitudes 33°06′ and 43°01′ north between longitudes 124°11; and 131°53′ east, in the northern temperate climate zone. Summers are hot and humid and winters severely cold. Rice is therefore a summer crop, grown between April and October. In the northern mountain regions, the rice plant can suffer from low temperatures at any stage between germination and maturity. In years of extreme low temperatures, all rice growing area as are susceptible to cold at the reproductive stage. For example, in 1980 and 1993, low temperatures damaged the Korean rice crop seriously, with grain yields dropping by 26% and 9.2%, respectively, compared with the national average yield on either side of these years (Fig. 42).

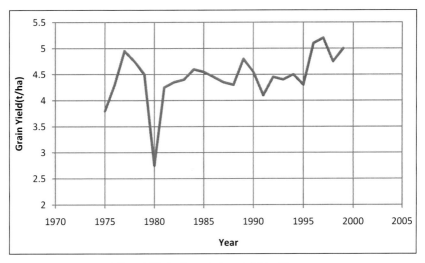

Fig. 42 Variation of rice yield in Korea.

Cold injury and damage are of varying degree and types mostly not known visibly, can only be analyzed. Some known injury is depicted hereunder (Table 31).

In the soil organic matter is depleting gradually thus application of organic matter is of most importance. Organic matter not only improves soil structure but also provides trace elements sufficiently apart from supply of major nutrients to the crops. Thus, in rice a combination beneficial effects are evident (Table 32).

Table 31 Type and symptoms of cold damage in rice

Growth stage	Critical temperature (°C)	Type of cold injury
Germination	10	Poor, delayed
Seedling	13	Retarded seedling growth Leaf discoloration, seedling rot
Vegetative	15	Inhibited rooting, growth and tillering Delayed panicle initiation
Reproductive	17	Inhibited panicle development Degenerated spikelets Disturbed meiosis and pollen formation Delayed heading
Heading	17	Poor panicle exsertion Inhibited anther dehiscence and pollination
Maturity	14	Poor grain filling and quality Early leaf senescence

Table 32 Effects of organic matter application on rice yield.

Treatment	Heading date in August	Filled grain ratio (%)	Brown rice yield (tha^{-1})
N-P-K (optimum)	12	10.3	0.44
50% + rice straw	10	58.1	2.33
50% + animal residue	9	49.8	2.36
50% + compost	9	52.5	2.31
50% + Compost + animal residue	9	54.2	2.58

In Faisalabad, Pakistan Akram et al. (2007) showed the yield of Basmati rices gradually decreases as the transplanting dates are delayed after July 10 due to occurrence of cold during reproductive phase. Yield reduction in delayed planting is attributed by decrease in panicles per hill and 1000 grain weight (Table 33).

Table 33 Yield and yield components as affected by date of transplanting.

Days of transplanting	Panicles per hill		Sterility (%)		1000 grain weight (g)		Paddy yield (tha⁻¹)	
	Basmati 385	Super Basmati	Basmati 385	Super Basmati	Basmati 385	Super Basmati	Basmati 385	Super Basmati
July 01	25.37	30.33	7.7	5.0	21.25	22.03	4.04	5.61
July 11	25.33	38.00	11.0	5.7	21.15	23.60	5.65	5.31
July 21	26.70	36.67	6.0	3.7	20.52	20.90	5.28	4.92
July 31	16.67	31.67	4.0	6.0	20.39	20.08	5.22	4.64
Mean	23.58	34.92	7.17	5.08	21.08	21.65	5.05	5.12

Thus, it is suggested that phosphate application has some beneficial effect on cultivation of rice in low temperature. Hence, judicious application of mineral nutrients specially nitrogen and phosphorus is important for higher yield even in abiotic stress conditions. Basuchaudhuri et al. (1986) studied with the rice genotypes from Kashmir, Bhutan, locals and Pusa-Khonorullo cross breeds at upper Shillong with cold stress at reproductive stage and described the agronomic traits. Khonorullo is a stable variety and the performances of other varieties are given in the Table 34. Minimum grain yield of 0.1t/ ha was noted in Barkat and a maximum grain yield of 2.5t/ha was recorded in PK-2-26-1.

It was also noted that suboptimal temperature at reproductive phase, along with neck blast and sheath rot are the major constraints to increase rice production in high altitude areas of NEH in India (Pandey et al. 1992) (Table 35).

Table 34 Agronomic traits of cold tolerant rices, Shillong.

Character	Range		Mean	Correlation
	Minimum	Maximum		Coefficient
				With grain
				Yield
Days to 50% flowering	121	170	159.31	0.224
	(K332)	(Khamang, Kba-sawrit, B2&B4)		
Plant height (cm)	60	110	88.77	0.477**
	(K332)	(Zichum)		
Leaf area index	1.84	6.70	3.27	0.143
	(Ryllored)	(PK-2-26-1)		
Tillers per plant	4	16	7.91	0.207
	(Kba-sawrit)	(PK-2-26-1)		
Effective tillers/plant	4	16	7.98	0.307*
	(Kba-sawrit)	(PK-2-26-1)		
Flag leaf angle (°)	70	115	93.27	0.387**
	(Namyi)	(PK-2-3-1)		
Panicle length (cm)	15	24	20.73	0.238
	(K332)	(Dull08, Ch1039)		
Grain per panicle	23	84.4	42.61	0.726**
	(Muzzudo)	(Ryllo red A)		
Grain Sterility (%)	27	98	52.30	−0.594**
	(Channa paddy)	(Muzzudo)		
Harvest index	0.01	0.70	0.35	0.602**
	(Borkat)	(Ryllored 6)		
Grain yield (t/ha)	0.1	2.5	1.91	-
	(Borkat)	(PK-2-26-1)		

It has been recognized that the maximum altitude in China for rice planting is 2710 m. Different climatic ecotypes of rice has different requirement of temperature condition.

Table 35 Yield of RCPL I-IC in upper Shillong. India, 1985–90.

Cultivar	Grain yield (t/ha)					
	1985	1986	1987	1988	1989	1990
RCPL I-IC	2.3	1.6	3.6	4.2	2.8	3.1
Khonorullo	1.9	1.3	2.9	3.3	2.1	2.5
Increase						
Over check (%) 22	25	24	28	25	16

Furthermore, geological environment and vertical decline rate of temperature vary from region to region. Therefore, rice has different maximum altitude in different regions. Planting rice in the mountain region with a large altitude difference, one can get a high productivity within the region. Research has indicated that the seed setting rate decreases with the increase in altitude (Yuan et al. 2005a). The reason is that the lower temperature restrains the transportation of content from the assimilate to grain which cannot be fully made use of (Liaud Lin 1990). As the influence of the altitude on the amount of grain and the effective panicles, an experiment in Mianyang, Sichuan province the altitude of which ranges from 400 m to 1400 m, has shown that the productivity with the increase of altitude, yield, the effective panicles and the quantity of grain first increased and then decreased with altitudes (Luo et al. 1999).

Planting rice in the mountain region with a large altitude difference, the rice can still get a high productivity within the safe stage if illumination, temperature and water are guaranteed, for, e.g., the average productivity of 'Shanyou 63' in Heenan province is merely 7t/ha. However, if it is transplanted to Taoyuan county (with an altitude of 1107 m) in Yunnan province, the productivity can reach to more than 15t/ha. In 2001, the researchers of Fujian Academy of Agricultural Sciences used the 'Teyon 175' and 'Ilyouming 86' in Taoyuan county which is the famous region of rice ultra high production making the rice yield records of 17.7t/ha and 17.9t/ha respectively. If the altitudes are different the average daily temperature of the whole growth period will be various, which will further influence the days of whole growth period and the seed setting rate decreases with the increase in altitude. The reason is that the low temperature restrains the transportation and content from the assimilate to grain which cannot be fully made use of. As to the influence of altitude on the grain number and the effective panicles, an experiment in Mianyang, Sichuan province where altitude ranges from 400 to 1400 m, has showed that the productivity with the increase of altitude, yield, the effective panicles and number of grains first increased and then decreased.

It has been reported that the quantity of grain of the same variety in Yunnan Taoyuan cultivated in Fujian Longlal, cultivated more than 50–60% area, and obtained a higher yield. This further showed that the effective panicles is the main reason of high yield about 'Shanyou 63' of Taoyuan county in Yunnan province the quantity of which is 59.9% higher than the average compared to Taoyuan county in Yunnan province (with an altitude of 1107 m) and Binchuan county (1438 m) and Hangzhou high yielding rice field differences. The results showed that compared with Hangzhou, Yunnan rice planting area increased blossom quantity mainly through the enhancement of the effective panicles and quantity of grain (Lang and Jichao 2012).

Studies on the phenology and yield activity of Basmati fine rice genotypes at Bangladesh, as influenced by planting date in monsoon season experienced a variation in yield. The late planting which experienced heat during grain filling produces less yield, thus suggesting a cold requirement for the crop (Mamun et al. 2009) (Fig. 43).

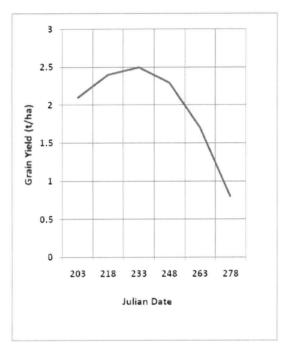

Fig. 43 Grain yield of basmati fine rice as affected by planting dates (aman season).

Late planted crop was affected adversely by low temperature, in Bangladesh, during panicle emergence and grain filling period resulting in lower grain yield (Salam et al. 1992).

It has also been noted that within a moderate range, low temperature during the reproductive stage appears to favour increasing spikelet number. In the northern region of Japan a large number of spikelets can be produced per unit nitrogen or dry weight of vegetative part than in southern Japan

(Murayama 1967). In other words, the efficiency of producing spikelets per unit nitrogen or dry weight is much higher in cool regions than in warm regions. In a controlled environment experiment where light intensity was kept high and plant nutrients were abundantly supplied lower temperature during the reproductive stage produced more spikelets (Yoshida 1973b).

Low temperature is a cause of sterility in cool regions.It is evident that air temperature as low as 15°C to 19°C at the meiotic stage of pollen mother cells formation causes very high sterility. This occurs in areas at high altitude in the temperate region and also at high altitudes of the tropics. Temperatures lower than the optimum may delay ripening and eventually fill grain percentage (Wada 1969).

In Ethiopia along the altitude at 3 places viz. Eladale (1616 masl, 28.4/12.4°C), Gomma^2 (1497 masl, 29.5/13.5°C) and Shebe (1420 masl, 30/14°C) 14 varieties of rice are being grown. Yield and yield components suggest that there is a drastic reduction in grain yield at Eladale along with the panicles per plant and grains per panicle as well as 1000 grain weight. However, grains per panicle are strongly and positively correlated with the grain yield (0.847) only (Tables 36 & 37).

Table 36 Yield and yield components in three locations in Ethiopia.

(Mean value)				
**				
Location	Panicles per plant	Grains per panicle	1000 grain wt (g)	Grain yield(Kg/ha)
**				
Eladale	6.01	22.49	23.82	749.98
Gomma 2	7.01	77.07	25.00	3092.38
Shebe	10.52	123.38	28.35	5184.44

Table 37 Correlations between grain field and yield components.

	Grains per panicle	1000grain wt (g)	Grain yield (Kg/ha)
Panicles per plant	0.632*	-0.139	0.497
Grains per panicle	1	-0.215	0.847**
1000 grain weight		1	-0.199
Grain yield			1

Spikelet fertility percentage was reduced with increased humidity also and with decreased temperature from 80 to 60% (Chunhoi and Zongtan 1990). Environmental yield potential is affected by climatic variation between sowing dates and years in high altitude. It is less influenced in mid altitude. Shrestha et al. (2012) has observed similar results in varying altitudes of Madagascar (Table 38).

Table 38 Grain yield (t/ha) at different altitude and different duration of genotype under cultivation of subsequent 2 years.

Grain yield (t/ha)					
	Early(1)	Early(2)	Late(1)	Late(2)	Mean
Mean of 10		High Altitude (1625 masl)			
Genotypes	2.10	3.10	1.10	2.70	2.25
		Medium Altitude (965masl)			
-do-	4.50	4.30	4.90	4.40	3.70
		Low Altitude (25 masl)			
-do-	2.80	3.10	2.00	2.70	2.65

Yield stability and genotype x environmental interactions of upland rice in altitudinal gradient of Madagascar suggests the importance of genotype -environment interaction (Fig. 44).

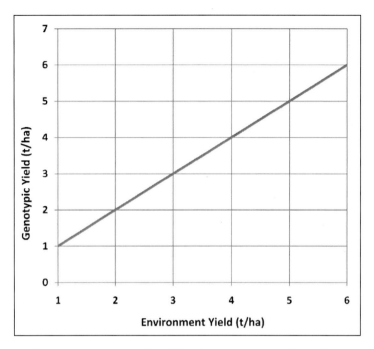

Fig. 44 Genotypic yield and Environment yield relationship.

An experiment to evaluate the grain yield potential of five temperate and fourteen tropical rice varieties grown under aerobic system in the dry season 2012 in the Ord River Irrigation Area (ORIA) indicated the yield variation in Australia. These varieties were selected based on their potential adaptation characteristic to tropical environment in Australia.

As it is true that the grain yield of rice under low temperature is negatively corrected with the cold stress score value during the growing period (Sivapalan 2013). Yunlu 29 is the only cold tolerant variety (Table 39 & Fig. 45).

In Turkey temperate region the association among yield and the yield components and their direct and indirect influence on the grain yield of rice were investigated. For this purpose, 80

Table 39 Average cold stress and grain yield.

Variety	Origin	Cold Stress Score	Grain Yield (t/ha)
		(Scale 0–10)	at 14% moisture
Yunlu 29	Yunnan , China	3.3	9.9
Langi	NSW , Australia	4.1	5.5
Viet5	Vietnam	4.7	4.9
Tachiminori	Japan	3.3	3.7
Viet 4	Vietnam	4.8	3.7
Fin	QLD , Australia	5.1	3.7
IR72	IRRI , Philippines	5.6	3.4
Kyeema	NSW , Australia	4.5	3.0
Doongara	NSW , Australia	4.1	2.8
Takanari	Japan	5.0	2.5
Quest	NSW , Australia	3.4	2.3
Illaboug	NSW , Australia	3.2	2.0
B6144	FMR-6 Indonesia	4.5	2.0
NTR587	IRRI , Philippines	6.4	2.0
Lemonet	Texas , USA	4.3	1.5
NTR426	IRRI , Philippines	8.6	1.0
IR64	IRRI , Philippines	5.8	0.9
Viet1	Vietnam	3.4	0.8
Pandan Wangi 7	Indonesia	7.7	0.8
	LSD5 %	0.04	3.76

breeding lines derived from 11 different cross populations in the F6 generations and their 10 parents were tested in a RCBD with two replications at the Thrace ARI in 1995. According to the results from the first year, 49 breeding lines were selected, and they and their 10 parents were tested in a RCBD in 3

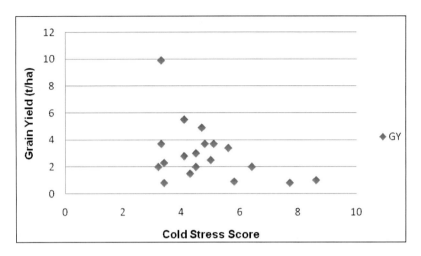

Fig. 45 Grain yield and cold stress score relation.

replications in the same institute in 1996. The phenotypic correlations among the traits and their path coefficients were estimated in both years.

Grain yield was significantly correlated with its components like number of productive tillers per m² (r=0.241** and r=0.274**), biological yield (r=0.803** and r=0.312**), harvest index (r=0.250** and r=0.677**) and number of field grains per panicle (r=0.496** and r=0.633**) in both years. Path coefficient analysis revealed that biological yield (0.748 and 0.481) and harvest index (0.413 and 0.704) had the highest positive direct effects on grain yield in both years. According to the magnitude of the direct effects on grain yield in both years the order of yield components was the number of filled grains per panicle (0.297 and 0.285)> the number of productive tillers per m² (0.233 and 0.197)> 1000 grain weight (0.165 and 0.136). The improvement in grain yield will be efficient, if the selection is based on the biological yield, the harvest index, the number of productive tillers per m² and number of filled grains per panicle under temperate conditions. However, both high biological yield and high harvest index should be taken into account together due to their negative correlations and indirect effects on each other (Surek and Beser 2003).

Gevrek (2012) noted the Aegean region with a typical Mediterranean climate on the West Coast of Anatolia, shows distinctive climate patterns different from any other area of the country. In this region, the lowermost and uppermost limits of temperature are 15°C–35°C respectively from beginning of April to the first week of september. Average rainfall is 500–600 mm. Hot and dry days of summer season is for 140 days from sowing to flowering time for rice cultivation in the region.

Minimum temperature variations.

Year	May	Jun	Jul	Aug	Sept	Oct
2003	13.5	19.6	20.3	21.4	16.3	13.7
2004	12.7	18.3	21.4	20.8	17.4	13.5

A varietal trial indicated that Demir is the highest yielding variety.

The analysis of yield and yield components is given here under (Table 40).

Experiment with elevated CO_2 and low temperature in rice at anthesis suggests that panicles are already formed, the effect is on the grain setting and grain filling process. Night temperature has a strong effect on grain set by increasing spikelet sterility, whereas elevated CO_2 has virtually no effect. Low temperature (22°C) has increased proportion of fertile spikelets. Grain weights have no significant difference along with brown rice yield (Cheng et al. 2009) (Table 41).

Table 40 Yield and yield components of rice in Turkey.

Variety	Yield (Kg/da)	Spikelets/main panicle (no)	Effective tillers/plant (no)	1000 grain weight(g)
Bal/Sariki/Kras – 424	528.0	445.0	7.4	29.0
Yavuz	625.0	515.0	7.4	30.3
Osmamick	732.0	562.0	7.2	33.2
Kiral	560.0	395.0	9.0	36.5
Negis	532.0	372.0	7.4	34.5
Denir	728.0	475.0	9.6	31.4
Kargi	588.00	397.0	8.1	36.2
Gonen	666.0	472.0	7.9	36.6
Toy/Arb/Nuc	616.0	467.0	7.4	32.4
Toy/So	576.0	467.0	8.7	30.1
Toag 92	496.0	355.0	8.4	33.2
Baldo	596.0	387.0	8.4	38.0
Mean	605.0	442.0	8.0	33.4
LSD (0.01)	58.68	29.15	0.48	1.50

Table 41 CO_2 and low temperature effects on yield and yield components.

Night temperature	CO$_2$ (abbrev.)	Panicle no. per hill	Spikelet no. per panicle	Fertile Spikelets (%)	Grain Wight (mg)	Brown rice Yield (g hill^{-1})	HI (%)
High (32°C)	Elevated (EH)	28.0	98.0	66.5	19.4	34.4	29.6
	Ambient (AH)	26.7	99.0	66.6	18.5	31.7	28.4
	% Change	5.0	−1.0	−1.0	5.1	8.5	4.2
Low (17°C)	Elevated (EH)	27.3	97.6	84.4	19.0	42.5	36.9
	Ambient (AH)	26.3	100.1	73.9	18.1	33.5	34.0
	% Change	3.8	−2.6	14.1	5.3	26.9	8.5

In China, it has shown that there was real linear relationship between light-temperature factors and grain yield. The size order of standard regression coefficient was: Crop illumination intensity > duration of day > average temperature > average illumination intensity > maximum temperature > effective accumulated temperature > minimum temperature. There was extremely significant negative correlation between average temperature and spike length while there was significant positive correlation between average illumination intensity and spike length. The particle correlation coefficient between average temperature and number of effective spikes per plant was bigger (r=0.4576). Except for duration of day, the correlation between other light-temperature factors and numbers of effective spikes per plant was significant. The partial correlation coefficient between duration of day and grains per panicle, seed setting rate were bigger which was 0.3746, –0.5599 respectively. The partial correlation coefficient between minimum temperature and 1000 grain weight was bigger (r=–0.1897) but the correlation between all the light-temperature factors and 1000 grain weight was not significant (Wen-Jiang and Wen-Yu 2010).

In Lao PDR, two varieties of rice grown during dry season of 2001–2002 at 5 different altitudinal heights showed some variations in grain yield (Sipasenth et al. 2007) (Table 42).

In one investigation in Bangladesh, genotypes having yield level more than 400 gm^{-2} were selected primarily as low temperature tolerant. There were short and long duration genotypes in those groups. The combination of yield components was quite better in groups. The genotypes having the highest spikelet number with better yield was found in group III and IV but they had lower 1000 grain weight. The maximum and minimum temperature around panicle

Table 42 Mean duration (days) from transplanting to harvest for variety TDK1 and TDK5 in the 2001–2002 dry season at each location.

Location	Altitude (m)	Mean Duration (Days)		Grain yield (t/ha)	
		TDK1	TDK5	TDK1	TDK5
Luang Namtha	560	131	124	5.58	5.53
Xieng Khoung	560	147	129	3.81	3.93
Luang Prabang	350	134	129	3.57	3.32
Sayatouli	290	107	97	3.48	2.91
Vientiane	171	108	101	4.08	4.01
Mean		125	116	4.10	3.94
LSD 5%		24		1.02	

initiation were 28.7°C and 10.9°C for short duration (SD) genotypes and 24.4 and 11°C for long duration (LD) genotypes. The genotypes encountered low temperature and low solar radiation from vegetative to reproductive phases. A few genotypes maintained better phenotypic acceptability. Most of the selected genotypes experienced spikelet degeneration symptoms. The genotypes (Group I, III and IV) producing huge number of spikelets might be able to execute panicle differentiation at the initial panicle development but at the subsequent stage there must have some growth impairment activities. In contrast, the genotypes (Group II and V) which having lower number of spikelets m^{-2} might be affected primarily at the differentiation stage, then at the division stage. The genotypes having short growth duration might have the tolerance to withstand low minimum temperature and those with long duration are appeared to have stress avoidance mechanism (Nahar et al. 2009) (Fig. 46).

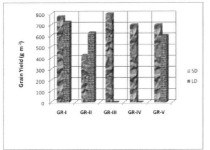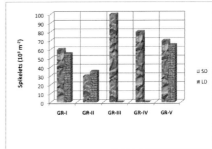
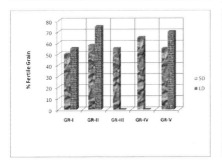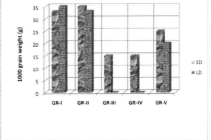

Fig. 46 Grain Yield, Spikelets per m^2, % fertile grains and 1000 grain weight variations under genotype groups.

Color image of this figure appears in the color plate section at the end of the book.

A close correlation between the cooling index and percentage of sterility was obtained from an analysis of spikelet sterility attributed to low temperatures at booting, that implies the day and night temperatures equally affect the incidence of spikelet sterility at booting (Fig. 47 & Table 43).

Thus it is evident that low temperature influenced the yield and yield components very much. But the degree of loss depends on the location specific situation and the other abiotic factors prevalent in the location. The span of cold spell is also important. Biotic factor associated with the low temperature also affects the yield. Hence, in totality it is quite complex and

145

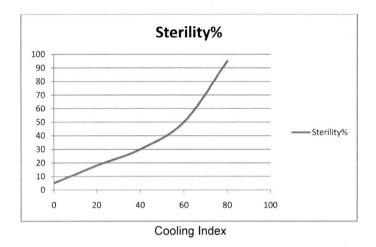

Fig. 47 Changes in sterility percent with cooling index.

Table 43 Effect of day and night temperatures on spikelet sterility at heading time of rice (after Shibata et al. 1970)D.

Duration of low temperature	Day temperature °C	Night temperature °C	Spikelet sterility(%)		Flowered spikelets(%) at heading
Days			Booting stage	Heading time	
	26	20	3.7	9.9	95
		14	6.4	19.4	36.7
		8	22.5	41.6	13.3
6	20	20	6.1	11.5	91.7
		14	6.6	15.5	18.3
		8	28.0	37.4	1.7
	14	20	7.0	14.6	38.3
		14	25.8	21.1	7.0
		8	74.8	48.3	1.3

as the yield and yield components respond appreciably under slight change of climatic change, it is important to understand the effects in manifestation of phenology in the rice plant.

7

CROP IMPROVEMENT

Rice is the major source of food for more than 2.7 billion people and planted on nearly one-tenth of earth's arable land. Of 130 million hectare of rice land 30% is subjected to salinity problems, 20% drought and 10% low temperature at high latitude and altitude areas were well documented in Japan (Shimono et al. 2007), Korea, North East and southern China, Bangladesh, India, Nepal and other countries (Lee 2001, Kaneda and Beachell 1974). In similar low temperature conditions severe yield loses were reported in Australia (Farrell et al. 2001), Italy and United States (Board et al. 1980).

Cold acclimatization is a process which increases the cold tolerance of the organism, after exposure to low, non freezing temperatures. The acclimatization ensures that cold tolerant species can endure harsh winter conditions, by preparing them to sub-zero temperatures, cold sensitive plants such as rice have limited abilities to cold climate and are therefore easily damaged during winter time.

The development of more tolerant varieties by using biotechnological methods is desirable, since the yields are expected to improve due to a prolonged vegetative period.

However, in order to apply such methods, more knowledge about the underlying mechanisms regulating the cold tolerance and acclimatization is required. EST sequencing and cDNA microarray technologies show that rice is cold responsive with many of the identified cold regulated genes having a counterpart in rice. In rice, however, the response is less dynamic than in the model organism Arabidopsis thaliana and this may explain its inability to fully acclimatize to cold. Identification of cis-elements coupled with transcription factors are prominent in the regulation of the response. Since cold acclimatization is a quantitative trait, the response of regulation of cold stress is under combinatorial control of several transcription factor and it is noted that this should be taken into account when identifying binding sites.

Many species of temperate origin may develop tolerance when exposed to temperature change. This process is known as thermal adaptation that is associated with biochemical and physiological responses caused mainly by alternations in lipidic fluidity of membranes (Hu et al. 2004). Cold acclimatization involves altered gene expression that affects membrane composition and accumulation of compatible solutes (Uemura et al. 2006). This is possible through the action of specific enzymes which are capable of altering the level of lipidic unsaturation of membranes. Therefore, fatty acid composition of the lipids that constitute the plant cell membranes. Genotypes differ for total saturated and unsaturated fatty acids only under the cold temperature treatment—more abundant fatty acids are linoleic, linolenic and palmitic which showed that the two differed between tolerant and sensitive genotypes. Linolenic acid increased after cold exposure in cold tolerant genotypes while palmitic acid decreased and an opposite behavior was found in the cold sensitive genotypes (Cruz et al. 2010).

A plant's ability to generate abundant energy becomes more important when it is put under additional stress, such as cold soil and air temperature.

Cold acclimatization of plants is a highly active process resulting from the expression of a number of physiological and metabolic adaptations to low temperature (Levitt 1980). Major metabolic changes in carbohydrates, proteins, nucleic acids, amino acids, and growth regulators have been documented during the acquisition of cold tolerance. Among these, water soluble carbohydrates like the fructose polymers and fructans were shown to accumulate during cold acclimatization of grass species (Pollock and Crains 1991, Livingston 1991). Fructans are claimed to enhance the cold tolerance in plants. The ability of plants to synthesize fructans correlates their survival in colder climates. Relationship between cold tolerance and fructan accumulation has been noted in cereals (Suzuki and Nass 1988, Pontis 1989). Evidence suggests that soluble sugars, such as sucrose and oligosaccharides of the raffinose family, in combination with heat stable proteins could play a determinant role in cold stress tolerance by protecting proteins and membranes against freeze-induced denaturation (Gusta et al. 1996).

Cold stress induced major changes in amino acid levels in wintering crowns of the three ecotypes and the highest contributions to total amino acid accumulation after acclimatization at low temperatures came from proline, glutamine and glutaric acid. Heat stable proteins have been isolated from cold acclimated plants and a correlation between a heat stable protein accumulation and cold induced freezing tolerance suggested that cold acclimatization induced proteins may act in combination with soluble carbohydrates and compounds in the acquisition of freezing tolerance (Gusta et al. 1996).

During cold acclimation, homologs of LEA (late embryogenesis abundant) proteins also accumulate in many plant species (Thomoshow 1999). During extracellular freezing, liquid water is withdrawn out of the cells, resulting in cellular dehydration (Guy 1990). Therefore, it has been suggested that LEA protein homologs may play a role in conferring tolerance in plant cells under freezing condition (Thomoshow 1999).

Cold acclimatization-induced accumulation of cold regulated (COR) proteins, the majority of these proteins have structural similarity with LEA proteins. LEA proteins have a highly hydrophilic feature and remain soluble upon boiling. LEA proteins have simple and a few amino acids especially hydrophilic proteins and share the unique property of heat stability.

Recently, to know the molecular basis, much emphasis has been given to gene expression. However, in order to apply such methods more knowledge and acclimatization is required. One step in this direction is to analyze gene expression data generated from cold stressed rice. EST sequencing and cDNA microarray technologies suggests that rice is cold responsive, to many of the previously identified cold regulated genes having their counterpart in the species. In rice, however, the response is less dynamic than in the model organization Arabidopsis thaliana and this may explain its inability to fully cold acclimatization (Lindlof 2008).

Plants respond with changes in their pattern of gene expression and protein products when exposed to low temperatures. Thus, ability to adapt has an impact on the distribution of survival of the plant, and on crop yields. Many species of tropical and subtropical origin are injured or killed by a nonfreezing low temperatures, and exhibit various symptoms of chilling injury such as chlorosis, necrosis or growth retardation. In contrast, chilling tolerant species are able to grow at such cold temperatures.

Conventional breeding methods have met with limited success in improving the cold tolerance of important crop plants involving inter-specific and inter-generic hybridization.

According to previous research, cold tolerance is controlled by many different genes. The mechanism of how these genes control cold tolerance is still not clear. Importance of cold tolerance by conventional breeding techniques is very difficult and time consuming. It requires expensive facilities for screening and it takes many breeding techniques to bring together all the important agronomic, physiological and quality traits. The solution would be marker assisted breeding, but first a better understanding of the molecular basis of cold tolerance is required.

In Chile there was breeding of 12 parents into the previously developed local population PQUI-I that segregates for a recessive male sterile gene, the parents are from different origins and represent a broad genetic base. Individual crosses were made between each parent and male sterile plants of PQUI-1. The F_1 seeds of each individual cross were mixed in different proportions and sown in Chile. F_2 seeds were sent to the International Centre for Tropical Agriculture (CIAT) in Colombia, for recombination, by harvesting seeds produced on the male sterile plants. Seeds from the first recombination were sent back to Chile for second recombination. The population PQUI-2, with broad genetic base represents a new starting point for temperate climate rice improvement through recurrent selection breeding and opens possibilities to breeders in the near future to use it as segregating population to derive and develop new promising lines and commercial varieties for the Chilean rice ecosystem (FAO 2003).

Lee (2001) mentioned a system of Korean rice breeding to make high yielding new cultivars with cold tolerance available (Fig. 48).

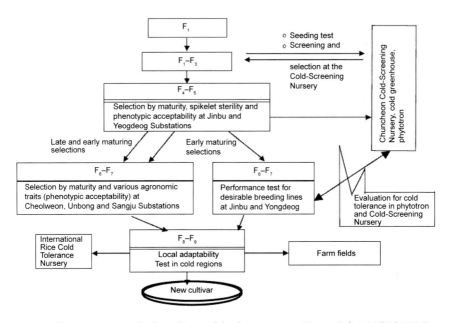

Fig. 48 Shuttle system for breeding cold tolerance rice, Korea (after NCES 1990).

Wide-compatibility varieties (WCVS) are a special class of rice germplasm that is able to produce fertile hybrids of rice germplasm, when crossed to both indica and japonica subspecies to be broken and provided the possibility of developing inter-subspecific hybrids in rice breeding programmes. However, the major problem in utilizing the japonica source of tolerance for developing improved cold tolerant indica rice cultivars is the high spikelet sterility in F1 and succeed in generations. Although, the degree and the frequency of partial sterility are overwhelming, rice breeding by means of wide crosses has been attempted and were successful in some cases a cold tolerant three way rice (*Oryza sativa* L.) cross involving indica, japonica and wide compatible variety (Purohit and Majumder 2009).

Recent studies involving full genome profiling/sequencing, mutational and transgenic plant analysis, have provided a deep insight of the complex transcriptional mechanism that operates under cold stress. The alterations in expression of genes in response to cold temperatures are followed by increases in the levels of hundreds of metabolites, some of which are known to have protective effects against the damaging effects of the cold stress. Various low temperature inducible genes have been isolated from plants. Most appear to be involved in tolerance to cold stress and the expression of some of them is regulated by c-repeat binding factor/dehydration responsive element binding transcription factors. Numerous physiological and molecular changes occur during cold acclimatization which reveals that the cold resistance is more complex than perceived and involves more than one pathway (Sanghera et al. 2011).

Rice seedlings are particularly sensitive to chilling in early spring in temperate and subtropical zones and in high elevation areas.

Improvement of chilling tolerance in rice may increase rice production significantly. MyBS3 is a single DNA binding repeat MyB transcription factor previously shown to mediate sugar signaling in rice. It was observed that MyBS3 also plays a critical role in cold adaptation in rice. Gain and loss of a function analysis indicated that MyBS3 was sufficient and necessary for enhancing cold tolerance in rice. Transgenic rice constitutively over expressing MyBS3 tolerated 4°C for at least 1 week and exhibited no yield penalty in normal field conditions. Transcription profiling of transgenic rice over expressing or under expression of MyBS3 led to the identification of many genes in the MyBS3 mediated cold signaling pathway. Several genes activated by MyBS3 as well as inducible by cold have previously been implicated in various abiotic stress responses and/or tolerance in rice and other plant species. Surprisingly, MyBS3 repressed the well known DREB1/CBF-dependent cold signaling pathway in rice, and the repression appears to act at the transcriptional level. DREB1 responded quickly and transiently while MyBS3 responded slowly to cold stress, which suggests that distinct pathways act sequentially complementarily for adapting short and long term cold stress in rice (Su et al. 2010). The distribution of low temperature germinability fitted a single gene segregation indicating that a single dominant gene with a large effect was transferred to the variety. This gene is tentatively symbolized as Ltg (t) (Fujino 2004).

The CBF/DREB1 genes represent one of the most significant discovery in the field of low temperature adaptation and signal transduction (Sanghera et al. 2011). Transgenic rice plants overexpressing Os DREB1 or At DREB1 genes slowed improved tolerance to low temperature, drought, and high salt stresses, and elevated contents of osmo protectants such as free proline and various soluble sugars. However, the transgenic plants show growth retardation under normal growth conditions (Its et al. 2006). Over expression of OsDREB1F gene also led to transgenic rice plants with enhanced stress tolerance, but no growth retardation effect was found, under normal growth conditions (Wang et al. 2008). Another rice DREB1 gene (OsDREB1D) was over expressed in Arabidopsis plants, resulting in transgenic plants in which the degree of cold tolerance was correlated with level of OsDREB1D expression (Zhang et al. 2009). OsDREB1D and OsDREB1A genes may be redundant in function (Zhang et al. 2009), since the level of cold tolerance that can be achieved by independent over expression of each one of them is very similar, along with the fact that both over expressed genes resulted in constitutive expression of COR15a, RD29A and KIN1, three genes involved in the plant cold tolerance (Dubouzet et al. 2003). Rice plants overexpressing the OsDREB1D gene have not been generated so far, but would probably have the same tolerant phenotype of rice plants over expression of other OsDREB1 genes. Recently Xu et al. (2011) overexpressed a maize CBF gene (ZmCBF3) in rice plants, and the resulting transgenics showed growth retardation only at the seedling stage, with no yield penalty under cold field conditions. As expected transgenic plants were cold tolerant (Xu et al. 2011).

Rice, a monocotyledonous plant that does not acclimatize to cold, has evidence differently from Arabidopsis, which acclimatizes to cold. To understand the stress response of rice in comparison with that of Arabidopsis, developed transgenic rice plants that constitutively expressed CBF3/DREB1A and ABF3, Arabidopsis genes that function in abscisic acid-independent and abscisic acid-dependent stress response pathways, respectively. CBF3 in transgenic rice elevated tolerance to drought and high salinity and produced relatively low levels of tolerance to low temperature exposure. ABF3 in transgenic rice increased tolerance to drought stress alone. By using the 60Kg rice whole genome microarray and RNA gel blot analysis identified 12 and 7 target genes that were activated in transgenic rice plants by CBF3 and ABF3, respectively, which appear to render the corresponding plants acclimatized for stress conditions. The target genes together with 13 and 27 additional genes are induced further upon exposure to drought stress, consequently making the transgenic plants more tolerant to stress conditions. Interestingly, transgenic plants exhibited neither inhibition nor visible phenotypic alterations despite constitutive expression of the CBF3 or ABF3, unlike the results previously obtained from Arbidopsis, where transgenic plants were stunted (Oh et al. 2005).

Panicle formation is interrupted by low temperature. It is generally accepted that cold tolerance of rice at one stage is different from another stage. However, Okabe and Toriyama (1972) reported that varieties seen to respond similarly to cold temperature at different growth stages. Some varieties have been found to be tolerant at different growth stages.

QTLs for cold tolerance related traits at the booting stage using balanced population for 1525 recombinant inbred lines of near isogenic lines (viz. NIL-RILs for BC5F3, BC5F4 and BC5F5) over 3 years and 2 locations by backcrossing the strongly cold-tolerant (Kunming X Iaobaigu) and a cold sensitive cultivar (Towada) was analyzed. 676 microsatellite markers were employed to identify QTLs conferring cold tolerance at booting stage. Single marker analysis revealed that 12 markers were associated with cold tolerance on chromosome 1, 4 and 5. Using a LOD significance threshold of 3.0 compositive interval mapping based on a mixed linear model revealed eight QTLs for 10 cold tolerance-related traits on chromosomes 1, 4 and 5. They were tentatively designated qCTB-1-1, qCTB-4-1, qCTB-4-2, qCTB-4-3, qCTB-4-4, qCTB-4-5, qCTB-4-6 and qCTB-5-1.

Their marker intervals were narrowed to 0.3–6.8 cM. Genetic distances between the peaks of the QTL and nearest markers varied from 0 to 0.04 cM. It is noticed in some traits associated with cold tolerance such as QCTB-1-1 for 5 traits (plant height, panicle exsertion, spike length, blighted grains per spike, and spikelet sterility), qCTB-4-1 for 8 traits (plant height, node length under spike, leaf length, leaf width, spike length, full grains per spike, total grains per spike and spikelet sterility), qCTB-4-2 for 3 traits (spike length, full grains per spike and spikelet fertility), qCTB-5-1 for 5 traits (plant weight, panicle exsertion, blighted grains per spike, full grains per spike and spikelet fertility). The variance explained by a single QTL ranged from 0.80 to 16.80%. Three QTLs (qCTB-1-1, qCTB-4-1 and qCTB-4-2) were detected in two or more traits. Then, it set a foundation for cloning cold-tolerance genes and provided opportunities to understand the mechanism of cold tolerance at the booting stage (Zeng et al. 2009).

Low temperature stress is common for rice grown in temperate regions and at high elevations in the tropics. The most sensitive stage of this stress is booting, about 11 days before heading. Japonica cultivars are known to be more tolerant than indicas. It had been constructed a genetic map using 191 recombinant inbred lines derived from a cross between a temperate japonica, M-202 and a tropical indica IR50 in order to locate qualitative trait loci (QTLs) conferring cold tolerance. The maps with a total length of 1,276.8 cM and an average density or one marker every 7.1 cM was developed from 181 loci produced by 175 microsatellite markers. Cold tolerance was measured as the degree of spikelet sterility of treated plants at a 12°C temperature for 5 days in the growth chamber. QTLs on chromosomes 1,2,3,5,6,7,9 and 12 were identified to confer cold tolerance at the booting stage. The QTL contribution to the phenotypic variation ranged from 11 to 17%. The two QTLs with the highest contribution to variation, designated qCTB 2a and qCTB3 were derived from the tolerant parent M-202, each explaining approximately 17% of the phenotypic variance. Two of the eight QTLs for cold tolerance were contributed by IR50.

In control condition, cold tolerance was measured as the percentage of reduction in panicle exsertion and in spikelet fertility. Evaluating cold tolerance through the reduction in panicle exsertion did not allow for the distinction between cold to tolerant from cold sensitive genotypes and when the spikelet fertility was considered, a minimum seven days was required to differentiate the genotypes for cold tolerance. Genotypes were more sensitive to cold at anthesis than at the microsporogenesis and as these stages were highly correlated, cold screening could be performed at anthesis only, since it is easier to determine. Cold tolerance of rice at the reproductive stage may be characterized by the reduction in spikelet fertility due to cold temperature (17°C) applied for seven days at anthesis (Cruz et al. 2010).

Further, it indicates the developmental regulation of the yield related genes pertaining to the genetic reprogramming involved at the corresponding developmental stage. The gene expression data can be utilized to specifically select particular genes which can potentially function synergistically for enhancing the yield while maintaining the source–sink balance. Furthermore, to gain some insight into the molecular basis of yield penalty during various abiotic stresses, the expression of selected yield-related genes has also been analyzed by QRT-FCR under such stress conditions. Analysis clearly showed a tight transcriptional regulation of a few of these yield-related genes by abiotic stresses. The stress responsive expression patterns of the genes could explain some of most important stress related physiological manifestations such as reduced tillering, smaller panicle and early completion of the life cycle causing reduced duration of vegetative and reproductive phases.

Development of high yielding rice varieties which maintain their yield even under stress conditions may be achieved by simultaneous genetic manipulation of certain combination of genes such as LRK1 and LOG, based on their function and expression profile obtained. In the future we will get to know whether over-expressing or knocking down such yield related genes can improve the grain yield potential in rice (Tripathi et al. 2012).

Abiotic stresses are serious threats to agriculture and natural status of the environment. Therefore, breeding for abiotic stress tolerance in crop plants and in forest trees should be given high research priority in plant biotechnology programmes. Molecular control mechanisms for abiotic stress tolerance are based on the activation and regulation of specific stress related genes. These genes are involved in the whole sequence of stress responses, such as signaling, transcriptional control, protection of membranes, and proteins, and free radical and toxic compound scavenging. Recently research into the molecular mechanisms of stress responses has started to bear fruit and, in parallel, genetic modification of stress tolerance has also shown promising results that may ultimately apply to agriculturally and ecologically important plants. Emphasis is placed on transgenic plants that have been engineered based on different stress response mechanisms specially in following aspects: regulatory controls, metabolite embryogenesis abundant and heat shock proteins (Fig. 49).

The rice cultivars released by research in Brazil are mostly belonging to the indica variety. They have high yield potential and grain quality, but are extremely sensitive to cold. In the state of Rio Grandedosul, the incidence of low temperatures during the early stage of development is one of the major limiting factors in rice productivity. About 100 indica genotypes were evaluated according to the survival percentage of plants at three leaf-stage after 10 days at 10°C (in a growth chamber) and seven days of recovery under normal temperature (in greenhouse conditions). The genotypes IRGA 959-1-2-2F-4-1-4-A and IRG A959-1-2-2F-4-1-1-D-1-CA-1 were characterized respectively as tolerant and susceptible to low temperature stress. Rice plants

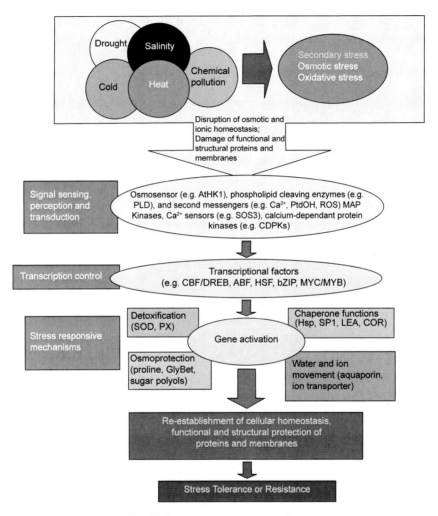

Fig. 49 Stress response mechanism.

Color image of this figure appears in the color plate section at the end of the book.

of the genotype, characterized as susceptible and tolerant to cold were maintained at 10°C for six hours. Specific primers of 14 rice genes expected to be involved in cold tolerance were designed to assess gene expression level by qRT-PCR. Tolerant and susceptible genotypes exhibited different levels of LIP9, WCOR413 and DREB1B gene expression. Expression of LIP was higher in the tolerant genotype after six hours of cold, whereas WCOR413 expression was highly inhibited by cold in both genotypes. DREB1B expression increased after six hours of cold mainly, in the tolerant genotype.

Low temperature is a common production constraint in rice cultivation in temperate zones and high elevation environments, with the potential to affect growth and development from germination to grain filling. There is a wide range of genotype based differences in cold tolerance among rice varieties, these differences often reflecting growth conditions in the place of origin, as well as breeding history. However, improving low temperature tolerance of varieties has been difficult, due to lack of clarity of the genetic basis to low temperature tolerance for different growth stages of the rice plant. Seeds or plants of 17 rice varieties of different origins were exposed to low temperature during germination (15°C), seedling, booting and flowering stages (18.5°C) to assess their cold tolerance at different growth stages. Low temperature at the germination stage reduced both the percentage and speed of germination. Varieties from China (B55, Benjiemang and Lijianghegu) and Hungary (HSC55) were more tolerant to low temperature than other varieties. Most of the varieties showed moderate levels of low temperature tolerance during the seedling stage, the exceptions being some varieties from Australia (Pelde, YRL 39 and YRM64) and Africa (WAB 160 and WAB38), which were susceptible to low temperature at the seedling stage. Low temperature at booting and flowering stages reduced plant growth and caused a significant decline in spikelet fertility. Some varieties from China (B55, Benjiemang, Lijianghegu), Japan (Jyoudeki), the USA (M103 and M104) and Australia (Quest) were tolerant or moderately tolerant while the remaining varieties were susceptible or moderately susceptible to low temperature at booting and flowering stages. These varieties from China (B55, Lijianghegu, Benjjiemang) and one from Hungary (HSC55) showed consistent tolerance to low temperature at all growth stages (Ye et al. 2012) (Tables 44 & 45).

Some rice varieties were searched in different countries to have local cold tolerant rice varieties. In the quest of that endeavour some rice varieties have been identified to be examined for cold tolerance throughout the world for further utilization.

Table 44 Cold tolerance of rice varieties at four growth stages (Ye et al. 2012 in press).

Variety	Origin	Germination	Seedling	Booting	Flowering
Amaroo	Australia	S	T	S	S
B55	China	T	T	T	T
Benjiemang	China	T	T	T	T
HSC 55	Hungary	T	T	T	T
Jyoudeki	Japan	S	T	M	M
Langi	Australia	M	M	M	S
Lijiangheigu	China	T	T	T	T
M103	USA	M	M	M	T
M104	USA	M	M	M	T
Quest	Australia	M	T	M	S
Reiziq	Australia	M	T	M	S
YRL39	Australia	S	S	S	T
YRM 64	Australia	S	S	M	S

S-Suseptible; M-Moderate; T-Tolerant

Table 45 Cold tolerant rice varieties

Variety	Country	Variety	Country
Longdao3	China	Longdao5	China
LonThur Hmwe	Myanmar	Pandan wangi	Indonesia
YRM 69	Australia	Amaroo	Australia
B55	China	Bengjiemang	China
Lijianghegu	China	HSC55	Hungary
Jyoudeki	Japan	M103	USA
M104	USA	Quest	Australia
Jinbubyco	Korea	Junganbyco	Korea
Quilla 64117	Chile	Osmaucik 97	Turkey
ESO76	Tanzania	Langi	Australia

Some popular rice varieties grown in high elevation of India and Nepal are HPU1, Kanchan, Himali, Himdhan, Himalaya1, K332, Kalimpong1, Khonorullo and Meghalaya1.

An early maturity indica rice variety Zhefu 49, whose grain quality and starch structure are sensitive to environmental temperature, was subjected to different temperature (32°C for high temperature and 22°C for optimum temperature) at grain filling stage in plant growth chambers, and the different expressions of their isoform genes (SBE1, SBEIII, SBEIV) encoding starch branching enzyme in the endosperms were studied by the real time fluorescence quantitative PCR (FQ-PCR) method. Effects of high temperature on the SBE-expression in developing rice endosperms were isoform-dependent. High temperature significantly down regulated the expressions of SBEI and SBEIII, while up regulated the expression of SBEIV. Compared with SBEIV and SBEIII, the expressions of SBEI in Zhefu 49 rice endosperms was more sensitive to temperature

165

variation at the grain filling stage thus, it indicates that changes in weather/climate conditions especially temperature stress influence rice formation and its quality as evident by isform expression (Wei et al. 2009). Although several japonica and some indica varieties of rice have already been transformed, there is significant scope for improvement in the technology of transformation of economically important indica varieties. Successful transformation of rice employing agrobacterium and recent advances in direct gene transfer by biolistics, evidenced by transfer of multiple genes, have removed some of the serious impediments in the area of gene engineering. The transfer of genes for nutritionally important biosynthetic pathway has provided many opportunities for performing metabolic engineering. Other useful genes for resistance against pests, diseases and abiotic stresses have also been transferred to rice. But the limited knowledge about important target genes require rapid progress in the field of functional genomics. The transgenic rice system can be applied to isolate new genes, promoters, and enhancers and their functions could be unraveled. The combination of novel regulatory systems for targeted expression and useful new genes should pave the way for improvement of rice.

REFERENCES

Abdellkhalik, A.F., E.M.R. Metwali, M. El-Adi, A.H. AbdEL-Hadi and D.E. EL-Sharnobi. 2010. Genotype-Environment interactions for seedling vigour traits in rice (*Oryza sativa* L.) genotypes grown under low and high temperature conditions. Am.-EUR. J. Agric. & Environ. Sci. 8: 257–267.

Akazawa, T. and I. Hara–Mishimura. 1985. Topographic aspects of photosynthesis, extracellular section and ultracellular storage of proteins in plant cells. Ann. Rev. Plant Physiol. 70: 441–472.

Akram, M., S.U. Ajmal and M. Munir. 2007. Inheritance of traits related to seedling vigour and grain yield in rice (*Oryza sativa*). Pak. J. Bot. 17: 74–77.

Alvarado, J.R. 1999. Influence of air temperature on rice population, length of period from sowing to flowering and spikelet sterility. In: Proceedings of Second Temperate Rice Conference pp. 63–68.

Alvarado, J.R. and P.B. Grau. 1991. Goiania Trabajos. Montevideo: IICA-PROCISUR pp. 105–114.

Andaya, V.C. and D.J. Mackill. 2003. QTLs conferring cold tolerance at the booting stage of rice using recombinant inbred lines from japonica x indica cross. Theor. Appl. Genet. 106: 1084–1090.

Anonymous. 1989. Annual Report, ICAR Research Complex for NEH Region, Shillong.

Ao, H., S. Wang, Y. Zou, S. Peng, Q. Tang, Y. Fang, Y. Chen, C. Xiong and A. Xiao. 2008. Study on yield stability and dry matter characteristics of super hybrid rice. Scientia Agricultura Sinica 41: 1927–1936.

Areunas–Huertero, F., A. Arroyo, L. Zhou, J. Sheen and P. Leon. 2000. Analysis of Arabidopsis glucose insensitive mutants. Genes and Development 14: 2085–2096.

Atkin, O.K., D. Bruhn, V.M. Hurry and M.G. Tjoelker. 2005. The hot and the cold unraveling the variable response of plant respiration to temperature. Func. Plant Biol. 32: 87–105.

Bano, A., K. Dorffling, D. Bettin and H. Hahn. 1993. Abscisic acid and cytokinins as possible root to shoot signals in xylem sap of rice plants in drying soil. Aust. J. Plant Physiol. 20: 109–115.

Basuchaudhuri, P., N.D. Majumder and D.N. Borthakur. 1986. Agronomic traits of cold-tolerant rices. IRRI Newsletter 11: 13–14.

Beck, E. and P. Ziegler. 1989. Biosynthesis and degradation of starch in higher plants. Ann. Rev. Plant Physiol. Plant Mol. Biol. 40: 95–117.

Bewley, J.D., M.J. Black and P. Halmer. 2006. In: The Encyclopedia of Seeds: Science, Technology and Use pp. 267–269.

Board, J.E., M.L. Peterson and E. Ng. 1980. Floret sterility in rice in a cool environment. Agron. J. 72: 483–487.

Bonneau, L., M. Carre and J. Martin-Tanguy. 1994. Polyamines and related enzymes in rice seeds differing in germination potential. Plant Growth Regulator 15: 75–82.

Bukhtoiarova, E.T., N.P. Krashook and J.A. Vishniakova. 1997. Changes in phosphate activity in rice seeds of varying germination. Prikl. Biokhim. Mikrobiol. 15: 494–498.

Champagne, E.T., D.F. Wood, B.O. Juliano and D.B. Bechtel. 2004. The rice grain and its gross composition. In: E.T. Champagne (ed.). Rice Chemistry and Technology, AACC Press, Mineapolis pp. 77–107.

Chamura, S. and M. Honma. 1973. Injury of low temperature to transplanted young seedling of rice plant (In Japanese). Niigata Agric. Sci. 25: 1–9.

Chen, P.W., C.M. Chiang, T.H. Tseng and S.M. Ye. 2006. Interaction between rice MYBGA and the Gibberettin response element controls tissue-specific sugar sensitivity of α-amylase genes. Plant Cell 18: 2326–234.

Chen, R., X. Zhao, Z. Shao, Z. Wei, Y. Wang, L. Zhu, J. Zhao, M. Sun, P. He and G. He. 2007. Rice UDP-Glucose phosphorylase 1 is essential for pollen callose deposition and its consuppression results in a new type of thermo sensitive genic male sterility. Plant Cell 19: 847–861.

Cheng, W., H. Sakai, K. Yagi and T. Hasegawa. 2009. Interactions of elevated CO_2 and night temperature on rice growth and yield. Agric. For. Met. 149: 51–58.

Cheng, W.H., A. Endo, L. Zhou, J. Penny and H.C. Chen. 2002. A unique short-chain dehydrogenase/reductase i glucose signaling and abscisic acid biosynthesis and functions. Plant Cell 14: 2723–2743.

Chung, G.S. 1979. The rice cold tolerance program in Korea. In: Rice cold tolerance workshop report, IRRI, Los Banos pp. 7–19.

Chungbuk, R.D.A. 1980. Annual Research Report, Cheongiu, Korea.

Chunhoi, S. and S. Zongtan. 1990. Spikelet fertility percentage as influenced by humidity and temperature. IRRI Newsletter 15: 10.

168

References

Cruz, L.J., G.B. Cagampang and B.O. Juliano. 1970. Biochemical factors affecting protein accumulation in the rice grain. Plant Physiol. 46: 743–747.

Cruz, R.P. and S.C.K. Milach. 2004. Cold temperature tolerance at the germination stage of rice: Methods of evaluation and characterization of genotypes. Sci. Agric. 61: 1–8.

Cruz, R.P., J.I. Golombieski, M. Cabreira, T.F. Silveira and L.P. Silva. 2010. Alternations in fatty acid composition due to cold exposure at the vegetative stage in rice. Brazilian Soc. Plant Physiol. 22: 199–207.

Cruz, R.P., S.C.K. Milach and L.C. Federizzi. 2006. Rice cold tolerance at reproductive stage in a controlled environment. Sci. Agric. 63: 255–261.

CSIRO. 2009. Annual Report, 2009.

Dai, L.Y., K. Kariya, C.R. Ye, K. Isc, Tannott, T.Q. Yu, F.R. Xu and C.W. Ma. 2002. Studies on the cold tolerance rice (*Oryza sativa* L.) II. South West China J. Agric. Sci. 15: 47–52.

Dai, Q. 1988. Effects of low temperature on the rice (*Oryza sativa* L.) root system at seedling stage. Philipp. J. Crop. Sci. 13: 9–13.

Dai, Q., J. Ziong, R.M. Visperas and B.S. Vergara. 1993. Response of rice (*Oryza sativa* L.) cultivars from China to low temperature stress. Philipp J. Crop Sci. 8: 99–106.

Dubouzet, J.G., Y. Sakuma, Y. Ito, M. Kasnga, E.G. Dubouzet and S. Mura. 2003. OsDREB genes in rice, *Oryza sativa* L., encode transcription activators that function in drought, high salt and cold responsive gene expression. Plant J. 33: 751–763 rice DREB1/CBF-type transc.

Dunayri, M., D.V. Tran and V.N. Nguyen. 1998. Reflections on yield gaps in rice production. How to narrow the gaps, FAO.

Emes, M.J., C.G. Bowsher, C.L. Hedley, M.M. Burrell, E.S.F. Scrase-Field and I.J. Tetlow. 2003. Starch synthesis and carbon partitioning in developing endosperm. J. Exp. Bot. 54: 569–575.

Engelke, T., J. Hirsche and T. Roitsch. 2010. Anther specific carbohydrate supply and restoration of metabolically engineered male sterility. J. Exp. Bot. 61: 2693–2706.

Engels, C., M. Hendrick, S. De Gryze and P. Tobback. 1986. Modelling water diffusion during long-grain rice soaking. J. Food Engineering 5: 55–73.

Fageria, N.K., A. Moreira and A.M. Coelho. 2011. Yield and yield components of upland rice as influenced by nitrogen sources. J. Plant Nutr. 34: 361–370.

Fairhurst, T.H. and A. Dobermann. 2002. Rice in the global food supply. In: Rice Production Spl. Supp. Publication, Better Crops Inter. 16: 3–7.

FAO. 2004. The state of food insecurity in the world, Rome p. 4.

FAO. 2010. FAOSTAT-Production Crops.

Farrell, T.C., K.M. Fox, R.L. Willams and S. Fukai. 2006. Genotypic variation for cold tolerance during reproductive development in rice: screening with cold air and cold water. Field Crops Res. 98: 178–194.

Farrell, T.C., K.M. Fox, R.L. Williams, R.E. Fukai and I.G. Lewan. 2001. In: Increased Low Land Rice Production in the Mekong Region. ACIAR, Canbera.

Ferrari, F., M. Fumagalli and A. Prolcini. 2009. Deciphering the proteomic profile of rice (*Oryza sativa*) bran: a pilot study. Electrophoresis 30: 4083–4094.

Fincher, G.B. 1989. Molecular and cellular biology associated with endosperm mobilization in germinating cereal grains. Ann. Rev. Plant Physiol. Plant Mol. Biol. 40: 305–345.

Finnie, C., S. Melchior, P. Roepstoff and B. Sevenson. 2002. Proteome analysis of grain filling and seed maturation in barley. Plant Physiol. 129: 1308–1319.

Fujino, K. 2004. A major gene for low temperature germinability in rice. Euphytica 136: 63–68.

Fujiwara, A. and H. Ishida. 1963. Nutriophysiological studies on the low temperature damaged rice plant. Part I. Influence of low temperature on the growth and nutrient absorption of rice plant in the vegetative growth stages (In Japanese). J. Sci. Soil Manure 34: 97–100.

Fu-rong, X., T.C. Feng, Y.T. Qiong, A.X. Xiang, Z.E. Lai, Y. Yuu, Z.D. Yu, D.X. Chao, P. Xi and D.L. Yuau. 2009. Genetic analysis of main characteristics related to cold tolerance in japonica rice from low latitude plateau and high latitude plain. Chinese J. Rice Sci. 28: 481–488.

Gao, J.P., D.T. Chao and H.X. Lin. 2008. Towards understanding molecular mechanisms of abiotic stress responses in rice. Rice 1: 36–51.

Garba, A., A.S. Fagum and G.G. Fuchison. 2007. Correlation of environmental factors to tillering and yield of rice during dry season in Banchi, Nigeria. IJOTAFS 1: 42–47.

Gevrek, M.N. 2012. Some agronomic and quality characteristics of new rice varieties in the Aegean region of Turkey. Turkish J. Field Crops 17: 74–77.

Gothandam, K.M., E. Nalini, S. Karthikeyan and J.S. Shiu. 2010. OsPRP$_3$, a flower specific proline rich protein of rice, determines extracellular matrix structure of floral organs and its over expression confers cold tolerance. Plant Mol. Biol. 72: 1`25–135.

Gubler, F., R. Kalla, J.K. Roberts and J.V. Jacobsen. 1995. Gibberellin regulated expression of a myb gene in barley aleurone cells. Evidence for myb transaction of a high p^{l}alpha amylase gene promoter. Plant Cell 7: 1879–1891.

Gunawardhana, T.A., S. Fukai and F.P.C. Blamey. 2003. Low temperature induced spikelet sterility in rice I. Nitrogen fertilization and sensitive reproductive period. Aust. J. Exp. Agric. 54: 937–946.

References

Guo-Li, W. and G. Zhen-fei. 2005. Effects of chilling stress on photosynthetic rate and chlorophyll fluorescence parameters in seedlings of two rice cultivars differing in cold tolerance. Rice Science 12: 187–191.

Gusta, L.V., R.W. Wilen and P. Fu. 1996. Low temperature stress tolerance: The role of abscisic acid, sugar and heat-stable proteins. Hort. Science 31: 39–46.

Guy, C.L. 1990. Cold acclimation and freezing stress tolerance: role of protein metabolism. Ann. Rev. Plant Physiol. Plant Mol. Biol. 41: 187–223.

Halford, N.G. and M.J. Paul. 2003. Carbon metabolite sensing and signaling. Plant Biotechnol. J. 1: 381–398.

Hamdani, A.R. 1979. In: Report of a Rice Cold Tolerance Workshop IRRI, Los Banos pp. 39–48.

Hansen, H. and K. Grossman. 2000. Auxin-induced ethylene triggers absciisic acid biosynthesis and growth inhibition. Plant Physiol. 124: 1437–1448.

Hassibi, P. 2010. Antioxidant responses of rice (*Oryza sativa* L.) seedings to low temperature stress. In: 3rd International Rice Congress, Hanoi, Vietnam.

Heenan, D.P. 1984. Low temperature induced floret sterility in the rice cultivars Calrose and Inga as influenced by nitrogen supply. Aust. J. Exp. Agric. Animal Husbandry 24: 255–259.

Herath, H.M. and D.P. Ormrod. 1965. Some effects of water temperature on the growth and development of rice seedlings. Agric. J. 57: 373–376.

Hirotsu, N., A. Makino, S. Yokota and M. Tadahiko. 2005. The photosynthetic properties of rice leaves treated with low temperature and high irridance. Plant Cell Physiol. 46: 1377–1383.

Hoshino, T., S. Matsushima, T. Tomita and T. Kukuchi. 1969. Analysis of yield determining process and its application to yield prediction and culture improvement of lowland rice. Proc. Crop. Sci. Soc. Japan.

Hu, Y.X., Y.H. Wang, X.F. Liu and J.Y. Li. 2004. Arabidopsis RAV1 is down regulated by brassinosteriod and may act as a negative regulator during plant development. Cell Res. 14: 8–15.

Ichishima, E. 1964. Protease activity in germinating rice seed. J. Ferm, Assoc. Japan 22: 393.

Igarashi, T. 2008. The influence of temperature during grain filling and the location of grains within rice (*Oryza sativa*) panicle on the amylose content of rice variety 'Kirava 397'. Jap. J. Crop Sci. 77: 142–150.

Its, Y., K. Katsura Maruyama, T. Taji, M. Kobayashi and M. Seki. 2006. Functional analysis of rice DREB1/CBF-type transcription factors involved in cold responsive gene expression in transgenic rice. Plant Cell Physiol. 47: 141–153.

Jackson, M.B., S.F. Young and K.C. Hall. 1988. Are roots a source of abscisic acid for the shoots of flooded pea plants. J. Exp. Bot. 39: 1631–1637.

171

Jacobs, B.C. and C.J. Pearson. 1994. Cold damage and development of rice: a conceptual model. Aust. J. Exp. Agric. 34: 917–919.

Jacobsen, J.V., F. Gubler and P.M. Chandler. 1995. Gibbrellin action in germinated cereal grains. In: P.J. Davis (ed.). Plant Hormones: Physiology and Biochemistry and Molecular Biology. Dordreht pp. 246–271.

Jagadish, S.V.K., P.Q. Craufurd and T.R. Wheeler. 2007. High temperature stress and spikelet fertility in rice (*Oryza sativa* L.). J. Exp. Bot. 57: 1627–1635.

Jie, H., L. Hong-xain, W. Yu-rou and J.K. Chun-yeu. 1981. Effects of low temperature on photosynthesis in flag leaves of rice (*Oryza sativa* L.). Acta Botanica Sinica 29.

Jin, Z., C. Qian, J. Ynag, H. Liu and Z. Piao. 2007. Changes in activities of GS during grain filling and their relation to rice quality. Rice Science 14: 211–216.

Kaneda, C. and H.M. Beachell. 1974. Response of indica-japonica rice to low temperature. SABRO J. 6: 17–32.

Kaplan, F., J. Kopka, D.W. Haskell, W. Zhao, K. Camerouschiller, N. Gatzke, D.Y. Sung and C.L. Guy. 2004. Exploring the temperature stress metabolome of Arabidopsis. Plant Physiol. 136: 4159–4168.

Kato, T., N. Sakurai and Kuraishis. 1993. The changes of endogenous abscisic acid in developing grains of two rice cultivars with different grain size. Jpn. J. Crop Sci. 62: 456–461.

Kaw, R.N. 1985. Characterization of the most popular rice varieties in low temperature areas of India and Nepal. Philipp. J. Crop. Sci. 10: 1–6.

Kazemitabar, S.K., A.B. Tomsett, M.C. Wilkinson and M.G. Jones. 2003. Effect of short term cold stress on rice seedlings. Euphytica 193–200.

Kharabian-Masouleh, A., D.L.E. Waters, R.F. Reinke, R. Ward and R.J. Henry. 2012. SNP in starch biosynthesis genes associated with nutritional and functional properties of rice. Scientific Reports 2, Article no. 557.

Kim, S.T., Y. Wang, S.Y. Kang, S.G. Kim, R. Rakwal and Y.C. Kim. 2009. Developing rice embryo proteomics reveals essential role for embryonic proteins in regulation of seed germination. J. Proteome Res. 8: 3598–3602.

Kishor, P.V.K., S. Sangam, R.N. Amrutha, P. SriLaxmi, K.R. Naidu, K.R.S.S. Rao, S. Rao, K.J. Reddy, P. Therippan and N. Srinivasulu. 2005. Curr. Sci. 88: 424–438.

Komatsu, S. and A. Kato. 1997. Varietal differences in protein phosphorylation during cold treatment of rice leaves. Phytochemistry 45: 329–335.

Kreps, J.A., Y. Wu, H.S. Chang, T. Zhu, X. Wang and J.F. Harper. 2002. Transcriptome changes for Arabidopsis in response to salt, osmotic and cold stress. Plant Physiol. 130: 2129–2141.

References

Kromer, S. 1995. Respiration during photosynthesis. Annu. Rev. Plant Physiol. Plant Mol. Biol. 46: 45–70.

Kuk, Y.I., J.S. Shin, N.R. Burgos, T.E. Hwag, O. Han, D.H. Cho and S. Jung. 2003. Antioxidative enzymes offer protection from chilling damage in rice plants. Crop Sci. 43: 2109–2117.

Kumagai, T., N. Arakai, Hamaoka and O. Ueno. 2011. Ammonia emission from rice leaves in relation to photorespiration and genotypic differences in glutamine synthetase activity. Ann. Bot. Pl. Prod. 14: 249–253.

Kurimoto, K., A.H. Millar, H. Lambers, D.A. Day and K. Noguchi. 2004. Maintenance of gowth rate at cold temperature in rice and wheat cultivars with a high degree of respiratory homeostasis iated with a high efficiency of respiratory ATP production. Plant Cell Physiol. 45: 1015–1022.

Kuwagata, T., J. Ishikawa-Sakurai, H. Hayashi, K. Nagasuga, K. Fukushi, A. Ahmad, K. Takasugi Kasuhata and M. Murai-Hatano. 2012. Influence of low air humidity and low root temperature on water uptake, growth and aquaporin expression in rice plants. Plant Cell Physiol. 53: 1418–1433.

Lang, L. and Y. Jichao. 2012. Research progress in effects of different altitude on rice yield and quantity in china. Greener J. Agric. Sci. 2: 340–344.

Lee, M.H. 2001. Low temperature tolerance in rice—the Korean experience, ACIAR, Canberra.

Lee, T.M. 1997. Polyamine regulation of growth and chililing tolerance of rice (*Oryza sativa* L.) root cultured *in vitro*. Plant Sci. 122: 111–117.

Leon, P. and J. Sheen. 2003. Sugar and hormone connections. Trends in Plant Science 8: 110–116.

Levitt, J. 1980. In: Chilling, Freezing and High Temperature Stress Vol.1. Responses of plants to environmental stress. Academic Press, New York.

Li, X., C.C. Dai, D.M. Jiao and C.H. Fover. 2006. Effects of low temperature in the light on antioxidant contents in rice (*Oryza sativa* L.) indica and japonica subspices seedlings. 32: 345–353.

Liang, C., L. Qiao-jun, S. Zong-xiu, X. Yong-zhong, Y. Xin–qiao and L. Li-jun. 2006. QTL mapping of low temperature on germination rate of rice. Rice Science 33: 93–98.

Liaud Lin. 1990. Final partitioning of the 14C fed before and after heading of paddy rice grown at different altitude localities. Acta Bot. Yunnanica 8: 459–466.

Lindlof, A. 2008. In the quest for a cold tolerant variety: gene expression profile analysis of cold stressed oat and rice. Thesis at Goteborgs Universitet p. 4.

Livingston, D.P. 1991. Non-structural carbohydrate accumulation in winter oat crowns before and during cold hardening. Crop Sci. 31: 751–755.

Li-Zhil, W., W. Chun-yan, L. Zhong-jiel, L. Ruil, L. Yu–yao, M. Ying and W. Lian-minl. 2009. Rice cooling injury in Heilongjiang Provience iv. Effect of low temperature on rice tillering. Heilongjiang Agri. Sci. 4.

Lu, B., Y. Yuan, C. Zhang, Jou, Y.W. Zhon and Q. Lin. 2005. Modulation of key enzyme involved in ammonium assimilation and carbon metabolism by low temperature in rice (*Oryza sativa* L.) roots. Plant Sci. 169: 295–302.

Luo, X., M. Zeng, Q. Zou and Z. Liang. 1999. Study on the changes in ecological environment and hybrid rice development at different altitudes in Sichuan. Chinese J. App. Environ. Biol. 12: 142–146.

Lyons, J.M. 1973. Chilling injury in plants. Ann. Rev. Plant Physiol. 24: 445–466.

Magor, N.P. 1984. A cropping pattern model for rainfed lowland rice in Bangladesh. M. Ag. Thesis, The University of Sydney. Australia pp. 3–38.

Mahajan, S. and N. Tuteja. 2005. Cold, salinity and drought stresses: An overview. Arch. Biochem. Biophy. 444: 139–158.

Majid, G.J., S. Ali and A.M. Seyed. 2011. Effects of the exogenous application of auxin and cytokinin on carbohydrate accumulation in grains of rice under salt stress. Plant Growth Regulator 65: 305–313.

Majora, D.J., W.M. Hammana and S.B. Rooda. 1982. Effects of short duration chilling temperature exposure on growth and development of shorghum. Field Crops Res. 5: 129–136.

Makino, A., T. Mae and K. Ohira. 1987. Variation in the contents and kinetic properties of ribulose-1, 5-bisphosphate carboxylases among rice species. Plant Cell Physiol. 28: 799–804.

Mamun, E.A., S. Alfred, L.C. Cantrill, R.L. Overall and B.S. Sutton. 2009. Effects of chilling on male gametophyte development in rice. Cell Biol. Inter. 30: 583–591.

Martin-Tanguy, J. 2001. Metabolism and function of polyamines in plant. Plant Growth Regulators 34: 135–148.

Maruyama, S., M. Yatomi and Y. Nakamura. 1990. Response of rice leaves to low temperature I. Changes in basic biochemical parameters. Plant and Cell Physiology 31: 303–309.

Massardo, F., L. Corcuera and M. Alberdi. 2000. Embryo physiological responses to cold by two cultivars of oat during germination. Crop. Sci. 40: 1694–1701.

Matsui, T. and H. Kagata. 2003. Characteristics of floral organs related to reliable self pollination in rice (*Oryza sativa* L.). Ann. Bot. 91: 473–477.

Matsui, T. and K. Omasa. 2002. Characteristics of floral organs related to self pollination in rice (*Oryza sativa* L.). Ann. Bot. 91: 473–477.

References

Matsui, T., K. Omasa and T. Horie. 1999. Rapid swelling of pollen grains in response to floret opening unfolds anther locules in rice (*Oryza sativa* L.). Plant Production Science 2: 196–199.

Matsushima, S., T. Tanaka and T. Hoshino. 1966. Analysis of yield determining process and its application to yield prediction and culture improvement of low land rice LXXV. Proc. Crop. Sci. Jpn. 34: 478–483.

Maxwell, P.H., M.S. Wiesener, G.W. Chang, S.C. Clifford, E.C. Vaux, M.E. Cockman, C.C. Wykoff, C.W. Pugh, E.R. Maher and P.J. Ratcliffe. 2009. The tumour suppressor protein VHL targets hypoxia-inducible factors for oxygen dependent proteolysis. Nature 399: 271–275.

Mayer, A.M. and A. Poljakoff-Mayber. 1989. The germination of seeds, Pergamon Press, Toronto pp. 174–178.

McDonald, D.J. 1979. Rice, Chapter 3 In: Australian Field Crops Vol. 2. Angus and Robertson, London pp. 70–94.

Millar, A.H., V. Miltova, G. Kiddle, J.L. Heazlewood, C.G. Bartoli, F.L. Theodoulou and C.H. Fover. 2003. Control of ascorbate synthesis by respiration and its implications for stress responses. Plant Physiol. 133: 443–447.

Mittler, R. 2002. Oxidative stress, antioxidants and stress tolerance. Trends in Plant Science 7: 405–410.

Mohapatra, P.K., R.K. Sarkar and S.R. Kumar. 2009. Starch synthesizing enzymes and sink strength of grains of contrasting rice cultivars. Plant Science 176: 258–263.

Morris, R.D., D.G. Blevins, J.T. Dietrich, R.C. Durley, S.B. Gelvin, J. Gray, N.G. Hommes, M. Kaminek, L.J. Mathews, R. Meilan, T.M. Reinbott and L. Sayavedra-Soto. 1993. Cytokinins in plant pathogenic bacteria and developing cereal grains. Aust. J. Plant Physiol. 20: 1547–1554.

Muntz, K. 1996. Proteases and proteolytic cleavage of storage proteins in developing and germinating dicotyledonous seeds. J. Exp. Bot. 47: 605–622.

Murayama, N. 1967. Nitrogen nutrition of rice plant. JARQ 2: 1–15.

Murty, K.S. 1980. In: Rice Cultivation in U.P. Himalaya Region: the present and future perspective, Wiley Eastern Limited p. 157.

Nagai, T. and A. Makino. 2009. Differences between rice and wheat in temperature responses of photosynthesis and plant growth. Plant Cell Physiol. 50: 744–755.

Nahar, K., J.K. Biswas, A.M.M. Shumsuzzaman and H.N. Barman. 2009. Screening of indica rice (*Oryza sativa* L.) genotypes against low temperature stress. Botany Res. Inter. 2: 295–303.

Nakamura, Y. 2002. Towards a better understanding of metabolic system for amylopectin biosynthesis in plants: rice endosperm as a model tissue. Plant Cell Physiol. 43: 718–725.

Ngchan, S.V., A.K. Mohanty and A. Pattanayak. 2010. In: Rice in North East India p. 35.

Nishiyama, I. 1976. Effect of temperature on the vegetative growth of rice plants. In: Proceedings of the Symposium on Climate and Rice. IRRI pp. 159–186.

Nishiyama, I. 1977. Physiology of cold injury in direct seeded rice with special reference to germination and seedling growth. Agric. Hortic. (Tokyo) 52: 1355–1357.

Nizigiyimana, A. 1990. Effect of low temperature on sterility of rice, Laboratorie Cytogenetique, Universite Catholique de Iouvain la neure, Belgium pp. 79–87.

Oh, S.J., S.J. Song, Y.S. Kim, H.J. Jang, S.Y. Kim, M. Kim, Y.K. Kim, B.H. Nahama and J.K. Kim. 2005. Arabidopsis CBF3/DREB1A and ABF3 in transgenic rice increased tolerance to abiotic stress without stunting growth. Plant Physiol. 138: 341–351.

Ohdan, T., P.B. Francisco, T. Sauda, T. Hirose, T. Terao, H. Satoh and Y. Nakamura. 2006. Expression profiling of genes involved in starch synthesis in sink and source organs of rice. J. Exp. Bot. 56: 3229–3244.

Okabe, S. and K. Toriyama. 1972. Tolerance to cold temperature Japanese rice varieties. In: Rice Breeding. IRRI p. 530.

Ortega, R.A., D.E. Delsolar and E. Acevedo. 2009. Spatial variability of spikelet sterility in temperate rice in Chile. EFITA conference p. 693.

Palmiano, E. and B.O. Juliano. 1972. Biochemical changes in rice grain during germination. Plant Physiol. 49: 751–756.

Palmiano, E. and B.O. Juliano. 1973. Changes in the activity of some hydrolases, peroxidase and catalase in rice seed during germination. Plant Physiol. 52: 274–277.

Pandey, D.K., H.S. Gupta, S. Kumar, G.C. Munda and M. Ram. 1992. RCPLI-IC, a cold-tolerant rice for high altitude areas of Meghalaya, India, IRRN 17: 14.

Parish, R.W., H.A. Phan, S. Lacuone and S.F. Li. 2012. Tapetal development and abiotic stress: a centre of vulnerability. Func. Pl. Biol. 39: 553–559.

Pathak, M.D. 1991. Rice cultivation in U.P. Himalayan region: The present and future perspective. In: Rice production in hilly eastern Uttar Pradesh p. 157.

Perata, P., C. Matsukura, P. Vemien and J. Yamaguchi. 1997. Sugar repression of gibberellins dependent signaling pathway in barley embryos. Plant Cell 9: 2197–2208.

Pollock, C.J. and A.J. Crains. 1991. Fructan metabolism in grasses and cereals. Ann. Rev. Plant Mol. Biol. 42: 77–101.

Pontis, H.G. 1989. Fructans and cold stress. J. Plant Physiol. 134: 148–150.

References

Purohit, S. and M.K. Majumder. 2009. A cold tolerant three way rice (*Oryza sativa* L.) cross involving indica, japonica and wide compatible variety. Middle-East J. Sci. Res. 4: 28–31.

Rao, S.P., B. Venkateswarlu and T.L. Acharyulu. 1984. Inter-relationships of grain size, number and spikelet filling for enhanced yield potential in rice. Indian J. Plant Physiol. 27: 281–289.

Richards, R.A. 2000. Selectable traits to increase crop photosynthesis and yield of grain. Crops 51: 447–458.

Rymen, B., F. Fabio Fatma, V. Klaas, I. Dirk, Gerritt and T.S. Beemster. 2007. Cold nights impair leaf growth and cell cycle progression in maize through transcriptional changes of cell cycle genes. Plant Physol. 143: 1429–1439.

Salam, M.A., M.A. Kabir and N.M. Miah. 1992. Adaptability of photoperiod sensitive modern rice varieties in boro season. Bangladesh Rice J. 3: 144–147.

Sanghera, G.S., S.H. Wani, Wttussaina and N.B. Singh. 2011. Engineering cold stress tolerance in crop plants. Current Genomics 12: 30–43.

Saruyama, H. and M. Tanida. 1995. Effect of chilling on activated oxygen-scavenging enzymes in the low temperature sensitive and tolerant cultivars of rice (*Oryza sativa* L.). Plant Sci. 109: 105–113.

Sasakawa, H. and Y. Yamntamoto. 1978. Comparison of the uptake of nitrate and ammonium by the rice seedlings: Influences of light, temperature, exogenous sucrose concentration, oxygen concentration and metabolic inhibitors. Plant Physiol. 62: 665–669.

Sasaki, T. 1968b. The relationship between germination under low temperature and subsequent early growth of seedling in rice varieties 1. On the elongation at early stage of seedling. Bull. Hokkaido Perfect Agric. Exp. Station. 17: 34–45.

Sasaki, T. and N. Yamazaki. 1971. The relationship between germination rate of rice seeds at low temperature and the subsequent early growth of seedlings IV on the establishment of seedlings. Proc. Crop. Sci. Japan 40: 474–479.

Satake, T. 1976. Determination of the most sensitive stage to sterile type cool injury in rice plants. Res. Bull. Hokkaido Natl. Agric. Exp. Stn. 113: 1–44.

Satake, T. and H. Hayase. 1970. Male sterility caused by cooling treatment at young microspose stage in rice plants V. Proc. Crop. Sci. Jpn. 39: 468–473.

Satya, P. and A. Saha. 2010. Screening for low temperature stress tolerance in boro rice. IRRN, 35.

Sen, K., M.M. Choudhuri and B. Ghosh. 1981. Changes in polyamine contents during development and germination of rice seeds. Phytochemistry 20: 631–633.

Shahi, B.B. and M.H. Heu. 1979. In: Report of a rice cold tolerance workshop, IRRI, Los Banos pp. 61–68.

177

Shahi, B.B., P. Whiteman and K.P. Shrestha. 1982. Strategies for development of cold tolerant rice varieties. SABRO J. 14: 143–152.

Shaw, J.F. and L.Y. Chuang. 1982. Studies on the -amylase from the germinated rice seeds. Bot. Bull. Academica Sinica 23: 45–61.

Shibata, M., K. Sasaki and Y. Shimazaki. 1970. Effect of air temperature and water temperature at each stage of the growth of lowland rice 1. Effect of air temperature and water temperature on the percentage of sterile grains. Proc. Crop Sci. Soc. Jpn. 39: 401–408.

Shieh, Y.J. and W.Y. Liao. 1987. Influence of growth temperature and nitrogen nutrition on photosynthesis and nitrogen metabolism in rice plant (*Oryza sativa* L.). Bot. Bull. Academia Sinica 28: 151–167.

Shih, F.F. 2004. Rice Proteins. In: Rice Chemistry and Technology, AACC International, St. Paul, MN.

Shimizu, M. 1958. Effects of temperature on the structure and physiology of the vegetative shoot apex in the rice plant. Japan J. Breed. 8: 195.

Shimono, H., M. Okada, E. Kanda and I. Arakawa. 2007. Low temperature induced sterility in rice. Evidence for the effects of temperature before panicle initiation. Field Crops Res. 101: 221–231.

Shinozaki, K. and E.S. Dennis. 2003. Cell signaling and gene regulation: Global analyses of signal transduction and gene expression profiles. Curr. Opin. In. Pl. Biol. 6: 405–409.

Shrestha, S.P., F. Asch, H. Breack, J. Dusserre and A. Ramantsouirina. 2012. Yield stability and genotype x environment interactions of upland rice in attitudinal gradient in Madagascar.

Sipasenth, S. Basnayke, S. Fukai, T.C. Farrell, M. Senthoughae, S. Engko, S. Phamixay, B. Li-quist and M. Charphengsay. 2007. Oppertunities to increasing dry season rice productivity in low temperature affected areas. Field Crops. Res. 102: 87–97.

Sivapalan, S. 2013. Cold tolerance of temperate and tropical rice varieties, WA Crop Updates.

Smeekers, S. 2000. Sugar induced signal transduction in plants. Ann. Rev. Plant Physiol. Plant Mol. Biol. 51: 49–81.

Smillie, R.M., R. Nott, S.E. Hetherington and G. Quist. 1987. Chilling injury and recovery in detached and attached leaves measured by chlorophyll fluorescence. Physiol. Plant. 69: 419–428.

Smyth, D.A. and H.E. Prescott Jr. 1989. Sugar content and activity of sucrose metabolism enzymes in milled rice grain. Plant Physiol. 89: 893–896.

Su, C.F., Y.C. Wang, C.A. Lu, T.H. Tseng and S.M. Yu. 2010. A novel $MyBS_3$-dependent pathway confers cold tolerance in rice. Plant Physiol. 63: 145–154.

References

Surek, H. and N. Beser. 2003. Correlation and path coefficient analysis for some yield related traits in rice (*Oryza sativa* L.) under Thrace conditions. Turkeish J. Agri. Forestry 27: 77–83.

Suzuki, M. and H.G. Nass. 1988. Fructan in winter wheat, triticle, and fall rye cultivars of varying cold hardness. Can. J. Bot. 66: 1723–1728.

Suzuki, S. 1982. Cold tolerance with rice plants with special reference to the floral characters II. Relations between floral characters and the degree of cold tolerance in segregating generations. Jap. J. Bread. 32: 9–16.

Takahashi, J., M. Yanagisawa, M. Kono, F. Yazawa and T. Yoshida. 1954. Studies on nutrient absorption by crops. Bull. Nat. Inst. Agric. Sci. B 4: 1–82.

Takahashi, N. 1984. Differentiation of ecotypes in *Oryza sativa* L. In: N. Takahashi and T. Sunoda (eds.). Biology of Rice. Tokyo, Japan pp. 31–67.

Tanaka, A. and T. Yamaguchi. 1969. Studies on the growth efficiency of crop plants (Part I). The growth efficiency during germination in the dark. J. Sci. Soil Manure 40: 38–42.

Tanaka, I. and S. Yoshitomi. 1973. Influence of low temperature in photosynthesis, photorespiration and transpiration of rice plant. Proc. Crop Sci. Soc. Japan 42: 109–110.

Tang, T., H. Xie, Y. Wang, B. Lu and J. Liang. 2009. 6th European Workshop on leaf senescence. J. Exp. Bot. 60: 2641–2651.

Tang, T., H. xie, Y. Wang, B. Lu and J. Liang. 2009. The effect of sucrore and abscisic acid interaction on sucrose synthase and its relationship to grain filling of rice (*Oryza sativa* L.). J. Exp. Bot. 60: 2641–2652.

Tewari, A.K. and B.C. Tripathy. 1998. Temperature stress induced impairment of chlorophyll biosynthetic reactions in cucumber and wheat. Plant Physiol. 117: 851–858.

Thomas, B.R. and R.L. Rodriguez. 1994. Metabolite signals regulate gene expression and source/sink relations in cereal seedlings. Plant Physiol. 106: 1235–1239.

Thomashow, M.F. 1999. Role of cold responsive genes in plant freezing tolerance. Plant Physiol. 118: 1.

Tripathi, A.K., A. Pareek, S.K. Soproy and S.L. Singhla Pareek. 2012. Narrowing down the targets for yield improvement in rice under normal and abiotic stress conditions via expression profiling of yield related genes. Rice 5: 37.

Tseng, T.H., C.S. Wang, C.L. Chen and J.M. Sung. 2003. Starch biosynthesizing enzymes in developing grains of rice cultivars. Tainung 67 and its sodium azide induced rice mutant. Field Crop Res. 84: 261–269.

Tsuneda, S. and A.H. Khan. 1968. Differences among strains of rice in the photosynthetic tissues I. A comperative leaf anatomy of indica and japonica. Tohoku J. Agric. Res. 19: 1–7.

Uemura, M., Y. Tominaga, C. Nakagawara, S. Shigematsu, A. Minami and Y. Kawamura. 2006. Responses of plasma membrane to low temperatures. Physiol. Plant. 126: 81–89.

Ueno, K. and M. Miyoshi. 2005. Difference of optimum germination temperature of seeds of intact and dehusked japonica rice during seed development. Euphytica 143: 271–275.

Wada, G. 1969. The effect of nitrogenous nutrition on the yield determining process of rice plant. Bull. Natl. Inst. Agric. Sci. Ser. A 16: 27–167.

Wang, Q., Y. Guan, Y. Wu, H. Chen, F. Chen and C. Chu. 2008. Overexpression of a rice OsDREB1F gene increases salt, drought and low temperature tolerance in both Arabidopsis and rice. Plant Mol. Biol. 67: 589–602.

Wei, K., F. Cheng, Q. Zhang and K. Liu. 2009. Temperature stress at grain filling stage mediates expression of three isoform genes encoding starch branching enzymes in rice endosperms. Rice Science 16: 187–193.

Weng, J.H. and C.Y. Chen. 1987. Differences between indica and japonica rice varieties in CO2 exchange rates in response to leaf nitrogen and temperature. Photosynthesis Res. 14: 171–178.

Wen-Jiang, L. and Y. Wen-Yu. 2010. Effects of light-temperature factors on grain yield and yield components in different latitude areas. Res. J. Agron. 4: 70–77.

Wer, F., H. Tao, S. Lin, B.A.M. Bruman, L. Zhang, P. Wang and K. Di Hert. 2011. Rate and duration of grain yield under different growing conditions. Sci. Asia 37: 98–104.

Xie, Z., D. Jiang, W. Cao, T. Dai and Q. Jing. 2003. Effects of post anthesis soil water status on grain starch and protein accumulation in specialty wheat varieties. J. Plant Physiol. Mol. Biol. 24: 309–316.

Xu, M., L. Li, Y. Fan, J. Wan and L. Wang. 2011. ZnCBF3 overexpression improves tolerance to abiotic stress in transgenic rice (*Oryza sativa*) without yield penalty. Plant Cell Rep. 30: 1949–1957.

Yamagata, H., T. Sugimoto, K. Tanaka and Z. Kasai. 1982. Biosynthesis of storage proteins in developing rice seeds. Plant Physiol. 70: 1094–1100.

Yamakawa, Y. and H. Kishikawa. 1957. On the effect of temperature upon the division and elongation of cells in the root of rice plant. Proc. Crop. Sci. Soc. Japan 26: 94–95.

Yamasaki, Y., S. Nakashimand and H. Komo. 2008. Pullanase from rice endosperm. Acta. Biochem. Pol. 53: 507–510 Cold-tolerant crop species have greater temperature homeostasis of leaf respiration and photosynthesis than cold-sensitive species. Plant Cell Physiol. 50: 203–215.

Yamori, W., J.R. Evans and S. Von Caemmerer. 2009. Effects of growth and measurement light intensities on temperature dependence of CO_2 assimilation rate in tobacco leaves. Plant Cell Environ. 33: 332–343.

References

Yang, J. and J. Zhang. 2013. Grain-filling problem in super rice. J. Exp. Bot. 61: 1–5.

Yang, J., J. Zhang, Z. Wang, L. Lin and Q. Zhu. 2003. Postanthesis water deficits enhance grain filling two-line hybrid rice. Crop. Sci. 43: 2099–2108.

Yang, J., J. Zhaug, Z. Wang, G. Xu and Q. Zhu. 2004. Activities of key enzymes in sucrose-to-starch conversion in wheat grains subjected to water deficit during grain filling. Plant Physiol. 135: 1621–1629.

Yang, P., X. Li, X. Wang, H. Chen, F. Chen and S. Shen. 2007. Proteomic analysis of rice (*Oryza sativa*) seeds during germination. Proteomics 18: 3358–3368.

Yatsuyanagi, S. 1960. Scheduled cultivation of rice in Tohoku region (in Japanese). Agric. Hortic. 35: 931–934.

Ye, N., G. Zhu, Y. Liu, A. Zhang, Y. Li, R. Liu, L. Shi, L. Jia and J. Zhang. 2011. Ascorbic acid and reactive oxygen species are involved in the inhibition of seed germination by abscisic acid in rice seeds. J. Exp. Bot. (online).

Ye, N., H. Zhu, X. Liu, A. Zhang, T. Li, R. Liu, L. Shi, L. Jia and J. Zhang. 2012. Ascorbic acid aive oxygen species are involved in the imbibitions of seed germination by absicsic acid in rice seeds. J. Exp. Bot. 63: 1809–1822.

Yoshida, H. and Y. Nagato. 2011. Flower development in rice. J. Exp. Bot. 62: 4719–4730.

Yoshida, S. 1973. Effects of temperature on growth of the rice plant (*Oryza sativa* L.) in a controlled environment. Soil. Sci. Plant Nutr. 19: 299–310.

Yoshida, S. 1973b. Effects of CO_2 enrichment at different stages of panicle development on yield components and yield of rice (*Oryza sativa* L.) temperature on growth of the rice plant. Soil. Sci. Plant Nutr. 19: 311–316.

Yoshida, S. 1978. Tropical climate and its influence on rice. IRRI Research Paper Series 20, Los Banos, Philippines.

Yoshida, S. 1981a. In: Fundamental of Rice Crop Science. IRRI, Los Banos pp. 1–63.

Yoshida, S. 1981b. Climate environment and its influence on the rice plant. In: Fundamentals of Rice Crop Science. IRRI, Los Bonos pp. 65–110.

Yu, S.M., Y.C. Lee, S.C. Fang, M.T. Chan, S.F. Hwa and L.F. Liu. 1996. Sugars act as a signal molecules and osmotic to regulate the expression of alpha a amylase genes and metabolic activities in germinating cereal grains. Plant Mol. Biol. 30: 1277–1289.

Yuan, J., Z. Ding, G. Cai, S. Yang, Q. Zhu and J. Yang. 2005a. The faction influencing RVA profile of rice starch and their changes with altitudes in Panxirecta Agron. Sinica 31: 1611–1619.

Zeng, Y., S. Yang, H. Cui, X. Yang, C. Xu, X. Pu, Z. Li, Z. Cheng and X. Huang. 2009. QTLs of cold tolerance-related traits at the booting stage for NIL-RILs in rice related by SSR. Genes and Genomics 31: 143–154.

Zeng-xun, J., Y. Jing, Q. Chun-rong, L. Hai-ying, J. Xue-yong and Q. Tai-quan. 2005. Effects of temperature during grain filling period on activities of key enzymes for starch synthesis and rice grain quality. Chinese J. Rice Sci. 19: 377–380.

Zhang, H., T. Chen, Z. Wang, J. Yang and J. Zhang. 2010. Involvement of cytokinins in the grain filling of rice under alternate wetting and drying irrigation. J. Exp. Bot. 61: 3719–3733.

Zhang, L., G. Zhao, J. Jia, X. Lia and X. Kong. 2012. Molecular characterization of 60 isolated wheat MYB genes and analysis of their expression during abiotic stress. J. Exp. Bot. 63: 203–204.

Zhang, Z.L., M. Shin, Z. Zou, J. Huang and T.H. Ho. 2009. A negative regulator encoded by a rice WEKY gene represses both absicsic acid and Gibberellins signaling in aleuronic cells. Plant Mol. Biol. 70: 139–161.

Zhao, J.Y. and Z.W. Yu. 2005. Effects of nitrogen fertilizer rate on nitrogen metabolism and protein synthesis of superior and inferior wheat kernel. Sci. Agric. Sinica 38: 1547–1554.

Zia, M.S., M. Salim, M. Aslam, M.A. Gill and Rahamatullah. 1994. Effect of low temperature of irrigation water on rice growth and nutrient uptake. J. Agron. and Crop. Sci. 173: 22–31.

INDEX

Color Plate Section

Chapter 3

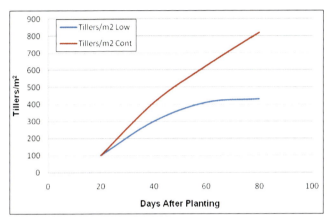

Fig. 13 Low temperature influence in tillering of rice plant.

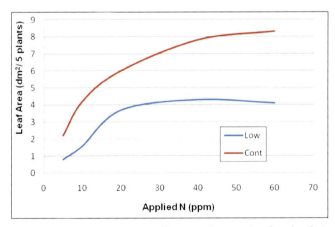

Fig. 14 Leaf area changes in rice seedlings under varying levels of nitrogen.

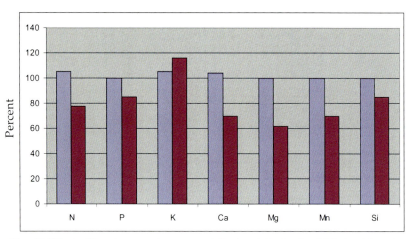

Fig. 19 Effect of low temperature on nutrient uptake in rice (Takahashi et al. 1954).

Color Plate Section

RuBPCase

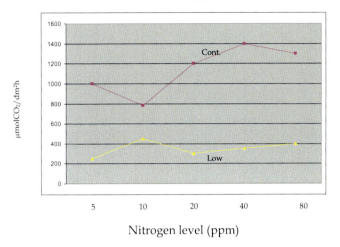

Nitrogen level (ppm)

Cytochrome C oxidase

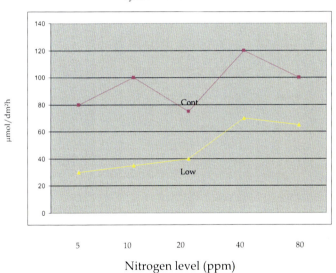

Nitrogen level (ppm)

Fig. 20 RuBPCase and Cytochrome C oxidase changes with temperature and nitrogen.

Chapter 4

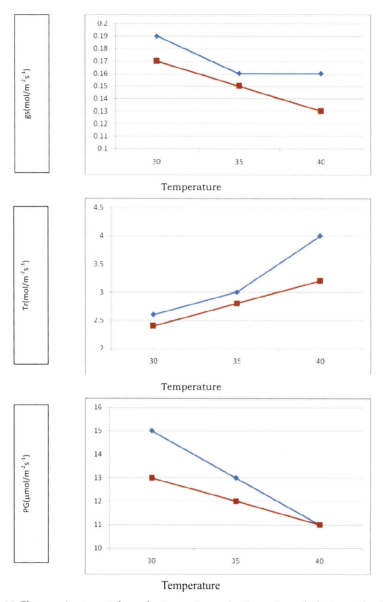

Fig. 29 Changes in stomatal conductance, transpiration rate and photosynthesis in leaves of rice genotypes at heading.

Chapter 5

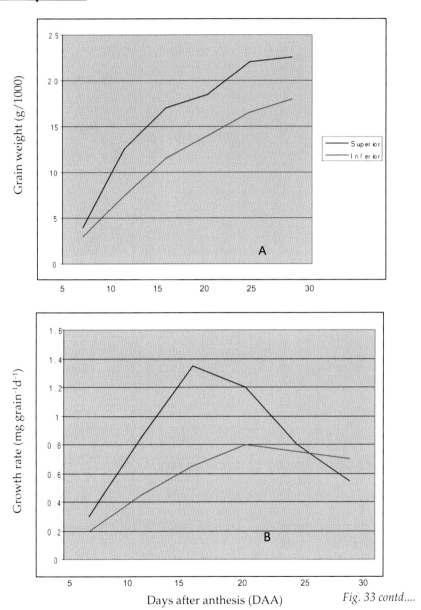

Fig. 33 contd....

Fig. 33 contd.

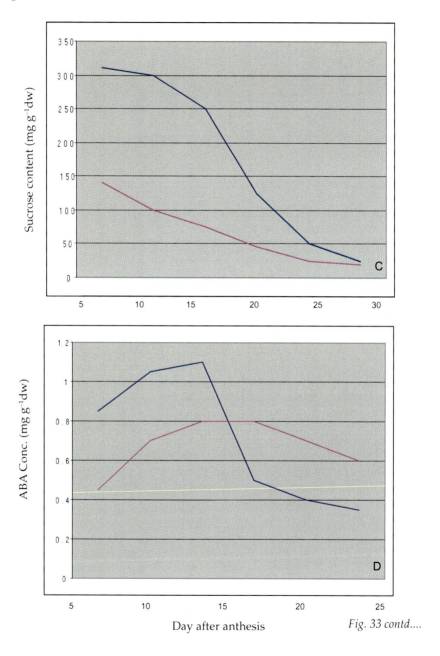

Fig. 33 contd....

Fig. 33 contd.

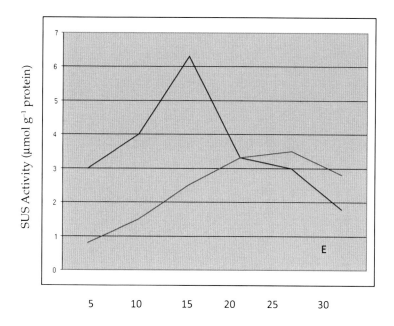

Fig. 33 Grain weight (A), Growth rate (B), Sucrose content (C), ABA Concentration (D) and SUS activity (E) changes during grain filling in superior and inferior grains.

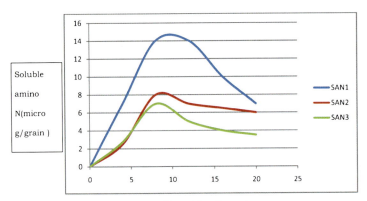

Soluble amino N(micro g/grain)

Days After Flowering

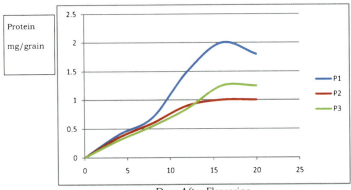

Protein mg/grain

Days After Flowering

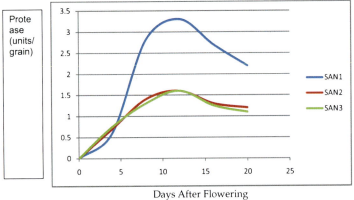

Protease (units/grain)

Days After Flowering

Fig. 37 contd....

Fig. 37 contd.

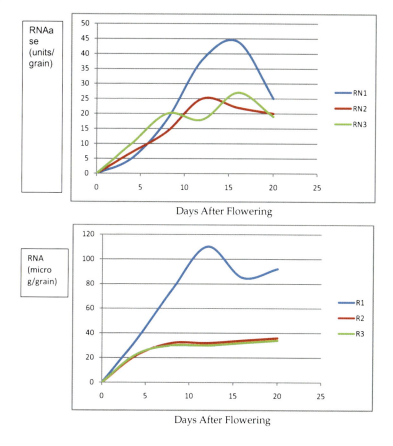

Fig. 37 Soluble amino N, Protein, Protease, RNase and RNA changes during grain filling in high protein and low protein rices.

Chapter 6

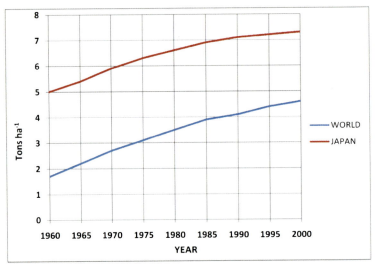

Fig. 39 Yearwise rice productivity in Japan and World.

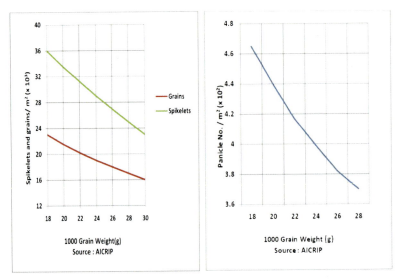

Fig. 40 Relationships of spikelets, grains and panicles per m² with 1000 grain wight.

198

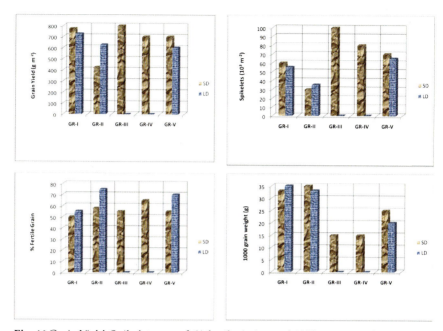

Fig. 46 Grain Yield, Spikelets per m², % fertile grains and 1000 grain weight variations under genotype groups.

Chapter 7

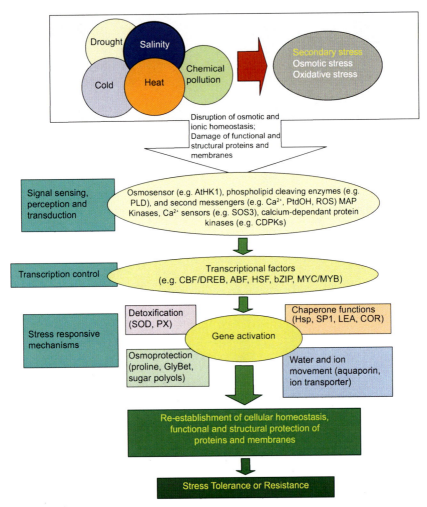

Fig. 49 Stress response mechanism.